PENGUIN BOOKS

MATHEMATICS: THE NEW GOLDEN AGE

Dr Keith Devlin has been a professional research mathematician since 1971, when he obtained his Ph.D. in mathematics from the University of Bristol. He has held positions at universities in England, Scotland, Norway, Germany, Canada and the USA. This book was commenced while he was Reader in Mathematics at the University of Lancaster in England, and first appeared during a two-year period in which he was a Visiting Associate Professor of Mathematics at Stanford University in California. He is currently the Carter Professor of Mathematics and Chair of the Mathematics Department at Colby College in Maine, USA.

In addition to his research work, since 1983 he has written a regular column on mathematics for the *Guardian*, has published occasional articles on mathematics in various scientific magazines, and has been involved in the production of a number of television programmes dealing with mathematical themes.

Keith Devlin

Mathematics:

The New Golden Age

PENGUIN BOOKS

PENGUIN BOOKS

Published by the Penguin Group
Penguin Books Ltd, 27 Wrights Lane, London W8 5TZ, England
Viking Penguin, a division of Penguin Books USA Inc.
375 Hudson Street, New York, New York 10014, USA
Penguin Books Australia Ltd, Ringwood, Victoria, Australia
Penguin Books Canada Ltd, 2801 John Street, Markham, Ontario, Canada L3R 1B4
Penguin Books (NZ) Ltd, 182–190 Wairau Road, Auckland 10, New Zealand

Penguin Books Ltd, Registered Offices: Harmondsworth, Middlesex, England

First published in Pelican Books 1988
Reprinted in Penguin Books 1990
10 9 8 7 6 5 4 3 2 1

Printed in England by Clays Ltd, St Ives plc
Typeset in 10/13pt Lasercomp Photina by
The Alden Press (London & Northampton) Ltd

Contents

Preface

The New Golden Age of Mathematics. When was the old one? Was it the period of the Ancient Greek geometers around 300 BC? Or did it occur during the seventeenth century when Newton and Leibniz were developing the infinitesimal calculus and Fermat was doing his tremendous work on number theory? Or perhaps the mathematical career of Gauss alone (1777–1855) justifies the title of Golden Age. Or, still later, the period that saw the work of Riemann, Poincaré, Hilbert, and others – the mathematics produced between the mid 1800s and the start of the Second World War was truly prodigious.

As with any area of human endeavour, it is not possible to say which was the truly 'greatest period'. Each new generation builds upon the work of its predecessors. What can be said is that the present time is witnessing the undertaking of a phenomenal amount of mathematical research. The *International Directory of Mathematicians* lists some 25 500 professional mathematicians around the world, but this represents only a small fraction of the real total. If you include also the vast armies of 'amateur mathematicians' (some of whom have made some significant discoveries!) for whom mathematics is simply a pleasant pastime, then the true figure must be enormous. On grounds of numbers (admittedly shaky grounds, since quantity and quality are not at all closely related, especially in mathematics), we are in the middle of a new Golden Age right now. And since every book has to have a title, that is more or less what I have chosen to call this one.

What this book sets out to do is to try to convey to the interested layperson some of the most significant developments that have taken place in mathematics during recent times. To include every advance that could be called 'significant' would require several volumes, not one. So I have had to be selective – very selective. First of all I restricted myself to developments which have taken place in the twenty-five years from 1960 to 1985, with a bias towards the latter part of this period. Since the book is intended for the general reader, I included only topics which have merited attention in the world's press and which are capable of explanation at a suitable level. And of course my own personal tastes and preferences played a role in my decisions.

For the most part, all that is required of you, the reader, is an interest in the subject that caused you to pick up the book in the first place, together with some degree of patience. (Understanding mathematics *takes time*, even at a superficial level.) There are, unavoidably, parts of the text where a reasonably good school mathematics education would enable you to get more from my account than would otherwise be possible, but I have tried to keep these to a minimum (and you can always skip over passages you find difficult, secure in the knowledge that it will soon get 'easier'). Though for the most part the chapters are quite independent of one another, they are ordered so that earlier ones might help in the appreciation of later ones.

Subject to all the above restrictions, plus the ever-present one of available space, I have tried to put across some of the richness and diversity of present-day mathematics. What you get is, I am afraid, just the tip of an iceberg. A book such as this is bound to fail in its aim; I only hope it does not fail too badly.

> Keith Devlin
> Lancaster, England
> May 1986

Acknowledgements

Like all mathematicians nowadays, I can call myself an expert in just one tiny area of a vast and growing landscape. In attempting to provide a comprehensive coverage, then, I have had to rely on others to pick up the errors that inevitably arose in my first draft. So thanks are due to: Sir Michael Atiyah, Amanda Chetwynd, David Nelson, Stephen Power, Hermann te Riele, Morwen Thistlethwaite, and David Towers, each of whom read all or part of the manuscript and made helpful suggestions. Thanks too to Penguin Books who from the very start showed great enthusiasm for what must have seemed like the impossible task of trying to produce a 'popular' account of mankind's most impenetrable subject. All failures and errors are, of course, to be laid at my door.

K. D.

For permission to reproduce copyright material grateful acknowledgement is made to the publishers listed below.

Figures 8 and 17–23 are reproduced from *The Beauty of Fractals* (1986), by H. O. Peitgen and P. H. Richter, with the permission of Springer-Verlag.

Figures 9 and 10 are reproduced from *Fractals: Form, Chance and Dimension* (1977), by Benoit Mandelbrot, with the permission of W. H. Freeman and Co.

Figure 14 is reproduced from *Studies in Geometry* (1970), by L. M. Blumenthal and K. Menger, with the permission of W. H. Freeman and Co.

Figures 31 and 34 are reproduced from *Scientific American* with the permission of W. H. Freeman and Co.

Figure 57 is reproduced with the permission of Cordon Art BV.

1 Prime Numbers, Factoring, and Secret Codes

The Biggest Prime Number in the World

The biggest (known) prime number* in the world is a giant which requires 65 050 digits to write out in standard decimal format. Using exponential (or power) notation it has a more manageable form:

$$2^{216091} - 1.$$

That is, you get the number by multiplying 2 by itself 216 090 times and then subtracting 1 from the answer.

Exponential notation is deceptive. To try to obtain some idea of its power for representing large numbers, imagine taking an ordinary 8 × 8 chessboard and placing piles of counters 2 mm thick (the English 10p piece is a good example) on the squares according to the following rule. Number the squares from 1 to 64, as in Figure 1. On the first square place 2 counters. On square 2 place 4 counters. On square 3 place 8 counters. And so on, on each square placing exactly twice as many counters as on the previous one. Thus on square n you will have a pile of 2^n counters. In particular, on the last square you will have a pile of 2^{64} counters. How high do you think this pile will be? 1 metre? 100 metres? A kilometre? Surely not! Well, believe it or not, your pile of counters will stretch out beyond the Moon (a mere 400 000 kilometres away) and the Sun (150 million kilometres away) and

*See later for an explanation of this term.

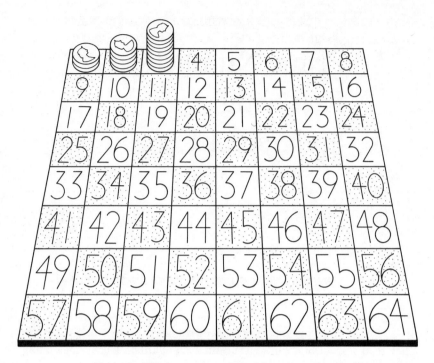

Figure 1. The astronomical chessboard number. By starting with two 2 mm thick coins on the first square and forming a pile twice as high on each successive square, the pile on the 64th square will stretch almost to the nearest star, Proxima Centauri, some 4 light years away.

will in fact reach almost to the nearest star, Proxima Centauri, some 4 light years from Earth. In decimal format the number 2^{64} is

$$18\,446\,744\,073\,709\,551\,616.$$

So much for 2^{64}. To obtain the number 2^{216091} which appears in the record prime expression you would need a chessboard with $216\,091$ squares – a board measuring 465×465 squares would do the trick!

Just how do you go about handling numbers of this size? For a start you use a computer. And not just any computer. The record prime mentioned above was discovered by using one of the most powerful computers in the world – a machine capable of performing two hundred billion* arithmetic

*Throughout this book 'billion' means a thousand million.

operations a second – and even then the calculation took over three hours. But computing power on its own is not enough; the skill of the mathematician is also required. How that skill was developed, and other uses to which it can be put, is the subject of the rest of this chapter.

Prime Numbers

'That action is best, which procures the greatest happiness for the greatest numbers', wrote Francis Hutcheson in 1725 (*Inquiry into the Original of our Ideas of Beauty and Virtue*, Treatise II, Section 3.8). It seems unlikely that he was thinking of *numbers* in the mathematical sense of greatest known primes and the like, but his statement nevertheless applies quite well to man's never-ending fascination with those most fundamental of mathematical objects – the *natural* (or *counting*) *numbers*, 1, 2, 3, ... These abstract mathematical objects are fundamental not only to our everyday life but to practically all of mathematics – so much so that the nineteenth-century mathematician Leopold Kronecker wrote (of mathematics): 'God created the natural numbers, and all the rest is the work of man.'

There are various properties which apply to natural numbers, splitting them into two classes (those with a property and those without). For instance, there is the property of being even. This splits the natural numbers into the class of those which *are* even (2, 4, 6, 8, ...) and those which are *not* (the odd numbers: 1, 3, 5, 7, ...). Or there is the property of being divisible by 3. (Here, as elsewhere in this book, when we say that one number *divides* another we mean that it does so exactly, leaving no remainder. Thus 3, 6, 9, 12 are all divisible by 3, whilst 1, 2, 4, 5, 7 are not.) The even–odd split is a natural and important one. (The split into those numbers divisible by 3 and those which are not is not so natural, nor of any great importance.) Another example of a natural and important split is given by the property of being a *perfect square*, like $1 = 1^2, 4 = 2^2, 9 = 3^2, 16, 25, 36, ...$. And there are many others. But by far the most important way of dividing up the natural numbers is into those which are *prime* and those which are not.

A natural number n is said to be a *prime number* if the only numbers which divide it are 1 and n itself. (The number 1 is a special case here, and it is conventional not to regard 1 as a prime number.)

Thus 2, 3, 5, 7, 11, 13, 17, 19 are all primes; 1, 4, 6, 8, 9, 10, 12, 14, 15, 16, 18, 20 are not. (Numbers which are not prime are sometimes called *composite*.) For instance, 7 is prime because none of the numbers 2, 3, 4, 5, 6 divides it; 14 is not prime since both 2 and 7 divide it.

The main reason why the prime numbers are so important was already known to the Greek mathematician Euclid (*c.* 350–300 BC) who, in Book IX of his *Elements* (a thirteen-volume compilation of all the mathematical knowledge then available) proved what is nowadays known as the *fundamental theorem of arithmetic*: that every natural number greater than 1 is either prime, or else can be expressed as a product of primes in a way which is unique except for the order in which the primes are arranged.

For instance, the number 75 900 is a product of seven *prime factors* (two being *repeated factors*):

$$75\,900 \;=\; 2 \times 2 \times 3 \times 5 \times 5 \times 11 \times 23.$$

The expression on the right of the equals sign here is called the *prime factorization* of the number 75 900.

What the fundamental theorem of arithmetic tells us is that the prime numbers are the basic building blocks from which all the natural numbers are constructed. As such they are like the chemist's elements or the physicist's fundamental particles. Knowledge of the prime factorization of any number provides the mathematician with almost complete information about that number, as is dramatically illustrated later on in this chapter (see the section on secret codes). But for now, what about the prime numbers themselves?

The most basic question you can ask about prime numbers is how common they are. Is there, for instance, a biggest prime number, or do the primes go on for ever, getting larger and larger? At first glance they seem to be very common indeed. Of the first ten numbers beyond 1 (i.e. 2 to 11 inclusive), five are prime, namely 2, 3, 5, 7, 11, which is exactly half the collection. Of the next ten numbers, 12 to 21, there are three which are prime (13, 17, 19), a proportion of 0·3. Between 22 and 31 the proportion of primes is again 0·3, whilst for the next two groups of ten numbers the proportion falls to 0·2. So the primes seem to 'thin out' the further you go along the sequence of natural numbers. Table 1 shows how the number of primes less than n (denoted by $\pi(n)$) varies with n for selected values of n, and gives the 'density' figure $\pi(n)/n$ in each case.

So, the primes become rarer the higher up you go in the number sequence. But do they eventually peter out altogether? The answer is no.

n	*π(n)*	*π(n)/n*
1000	168	0·168
10000	1229	0·123
100000	9592	0·096
1000000	78498	0·078

Table 1. The distribution of primes, showing the number of primes $\pi(n)$ smaller than *n* for various values of *n*.

This was also demonstrated by Euclid, using an argument which to this day remains a superb model of elegant mathematical reasoning. To begin with, imagine the prime numbers listed in order of magnitude:

$$p_1, p_2, p_3, \ldots .$$

So $p_1 = 2, p_2 = 3, p_3 = 5$, and so on. The aim is to show that this list must continue for ever. To put it another way, it has to be demonstrated that if we are at any stage *n* in the list, having enumerated $p_1, p_2, \ldots , p_n$, then there has to be a further prime in the list beyond p_n. The trick is to look at the number

$$N = p_1 p_2 p_3 \cdots p_n + 1$$

obtained by multiplying together all the primes p_1, p_2, p_3 and so on up to p_n, and then adding 1 to the result. Obviously *N* is bigger than p_n, so if *N* happens to be prime then we know that there is a prime beyond p_n, which is what we are trying to prove. On the other hand, if *N* is not prime it must be divisible by some prime, call it *p*. But if you try to divide *N* by any of the primes $p_1, p_2, \ldots , p_n$ there will be a remainder of 1 (the same 1 that was added when we obtained *N* in the first place). So our *p* must be a different prime, and again we have proved what was required. So, in any event there will be a prime bigger than p_n, and we can conclude that the list of primes continues for ever.

Notice that we have no idea whether the number *N* in the above is prime or not. If you try a few examples you will discover that numbers of this form often are prime. For instance,

$$N_1 = 2 + 1 = 3,$$
$$N_2 = 2 \times 3 + 1 = 7,$$
$$N_3 = 2 \times 3 \times 5 + 1 = 31,$$
$$N_4 = 2 \times 3 \times 5 \times 7 + 1 = 211,$$
$$N_5 = 2 \times 3 \times 5 \times 7 \times 11 + 1 = 2311,$$

are all primes. But the next three are not:

$$N_6 = 2 \times 3 \times 5 \times 7 \times 11 \times 13 + 1$$
$$= 30031 = 59 \times 509,$$
$$N_7 = 19 \times 97 \times 277,$$
$$N_8 = 347 \times 27953.$$

In fact, no one knows whether infinitely many numbers of the form

$$N_n = p_1 p_2 \cdots p_n + 1$$

are prime, nor indeed whether infinitely many of these numbers are composite (though at least one of these two possibilities has to be true, of course). This is just one of dozens of easily stated questions about prime numbers whose answer is not known.

One of the most famous unanswered questions about prime numbers is *Goldbach's conjecture*. In a letter to Leonhard Euler written in 1742, Christian Goldbach conjectured that every even number greater than 2 is a sum of two primes. For instance,

$$4 = 2 + 2,$$
$$6 = 3 + 3,$$
$$8 = 3 + 5,$$
$$10 = 5 + 5,$$
$$12 = 5 + 7.$$

Computer searches have verified Goldbach's conjecture for all even numbers up to 100 000 000, but to this day the conjecture has not been settled properly one way or the other.

Primality Testing

Though most of the classical problems concerning prime numbers have remained unsolved, the last few years have seen tremendous developments in methods by which numbers may be tested to see if they are prime or not. 'Methods for testing primality?' you cry. 'But surely it is obvious how to go about it?' And indeed, there is a perfectly natural, straightforward way of seeing if a number is prime or not. Given your number, n say, you first see if 2 divides it. If it does, then n is not prime and that is the end of the matter. Then you try 3. If 3 divides n, then n is not prime and again you are finished. Then try to divide n by 5. (You can skip past 4: since 2 does not divide n if you have got this far, 4 cannot divide n either.) If 5 fails to divide n, you try 7. (Again, you can skip past 6 since 2 and 3 do not divide n.) And so on. If you get as far as $\sqrt{n}$ without finding a number which divides n, then you know that n must be prime. (Because if n were not prime it would be a product of two numbers u and v between 1 and n, and either u or v will be no greater than $\sqrt{n}$, of course.)

The above process is known as *trial division*. Though it works well enough for moderately small numbers, it becomes unwieldy if the numbers are too large. To see just how impractical it becomes, suppose you were to write a highly efficient program to run trial division on the fastest computer available (alluded to at the start of this chapter). For a number of 10 digits the program would appear to run instantaneously – the answer would flash up immediately. For a 20-digit number it has a bit of a struggle and would take two hours. For a 50-digit number it would require a staggering ten billion years. A 100-digit number would require this many years:

$$1\,000\,000\,000\,000\,000\,000\,000\,000\,000\,000\,000\,000$$

(there are thirty-six zeros here). This is not just a trivial calculation of a very large number. As will be explained later in this chapter, primes with between 60 and 100 digits are required for one of the most secure forms of secret coding system in use today.

So just how do you go about deciding whether a 100-digit number is prime? The best method available at the moment is a highly sophisticated technique developed around 1980 by the mathematicians Adleman, Rumely, Cohen, and Lenstra, and often referred to by their initials as the ARCL test. When implemented on the same type of computer as mentioned above, the running times for the ARCL test are, for a 20-digit number, 10 seconds; for a 50-digit number, 15 seconds; for a 100-digit number, 40 seconds. The computer will even handle a 1000-digit number if you give it a week to work on the problem.

How does the test work? Well, it depends on a considerable amount of highly sophisticated mathematics – mathematics way beyond a typical undergraduate degree course – so it is not possible to give a complete answer here. But it is not hard to explain the central idea behind the method. This is a simple (though very clever) piece of mathematics due to the great French mathematician Pierre de Fermat (1601–65).

Though only an 'amateur' mathematician (he was a jurist by profession), Fermat produced some of the cleverest results mathematics has ever seen, even to this day. One of his observations was that if p is a prime number, then for any number a less than p, the number $a^{p-1} - 1$ is divisible by p. For instance, suppose we take $p = 7$ and $a = 2$. Then

$$a^{p-1} - 1 = 2^6 - 1 = 64 - 1 = 63,$$

and indeed 63 is divisible by 7. Try it yourself for any values of p (prime) and a (less than p). The result is always the same.

So here is a possible way of testing if a number n is prime or not. Compute the number $2^{n-1} - 1$. See if n divides it. If it does not, then n cannot be prime. (Because if n *were* prime, then by Fermat's observation you *would* have divisibility of $2^{n-1} - 1$ by n.) But what can you conclude if you find that n does divide $2^{n-1} - 1$? Not, unfortunately, that n has to be prime. (Though this is quite likely to be the case.) The trouble is, whilst Fermat's result tells us that n divides $2^{n-1} - 1$ whenever n is prime, it does not say that there are no composite numbers with the same property. (It is like saying that all motor cars have wheels; this does not prevent other things having wheels – bicycles, for instance.) And in fact there are non-primes which do have the Fermat property. The smallest one is 341, which is not prime since it is the product of 11 and 31. But if you were to check (on a computer) you would find that 341 does divide $2^{340} - 1$. (We shall see in a moment that there is no need to calculate 2^{340} in making this check.) Composite numbers which behave like primes as far as the Fermat property is concerned are called *pseudoprimes*. So if, when you test for primality using the Fermat result, you discover that n does divide $2^{n-1} - 1$, then all you can conclude is that either n is prime or else it is pseudoprime. (In this case the odds are heavily in favour of n actually being prime. For though there are in fact an infinity of pseudoprimes, they occur much less frequently than the real primes. For instance there are only two such numbers less than 1000, and only 245 below one million.)

Incidentally, it makes little difference if instead of 2 you use some other number, say 3 or 5, in testing the Fermat property. Whichever number you use there will be pseudoprimes to prevent you obtaining an absolute answer to your primality problem.

In using the above test, it is not necessary to calculate the number 2^{n-1}, a number which we have already observed to be very large for even quite modest values of n. All you need to do is find out whether or not n divides $2^{n-1} - 1$. This means that multiples of n may be ignored *at any stage of the calculation.* To put it another way, what has to be calculated is the remainder that *would* be left *if* $2^{n-1} - 1$ *were* divided by n. The aim is to see if this remainder is zero or not, but since multiples of n will obviously not affect the remainder, they may be ignored. Mathematicians (and computer programmers) have a standard way of denoting remainders: the remainder left when A is divided by B is written as

$$A \bmod B.$$

Thus, for example, $5 \bmod 2$ is 1, $7 \bmod 4$ is 3, and $8 \bmod 4 = 0$.

As an example of the Fermat test, let us apply it to test the number 61 for primality. We need to calculate the number

$$(2^{60} - 1) \bmod 61.$$

If this is not zero, then 61 is not a prime. If it is zero, then 61 is either a prime or a pseudoprime (and in fact is a genuine prime, as we know already). We shall try to avoid calculating the large number 2^{60}. We start with the observation that $2^6 = 64$, and hence $2^6 \bmod 61 = 3$. Then, since $2^{30} = (2^6)^5$, we get

$$2^{30} \bmod 61 = (2^6 \bmod 61)^5 \bmod 61 = 3^5 \bmod 61$$
$$= 243 \bmod 61 = 60.$$

So,

$$2^{60} \bmod 61 \;=\; (2^{30})^2 \bmod 61 \;=\; (2^{30} \bmod 61)^2 \bmod 61$$
$$=\; 60^2 \bmod 61 \;=\; 3600 \bmod 61 \;=\; 1.$$

Thus,

$$(2^{60} - 1) \bmod 61 \;=\; 0.$$

Since the final answer here is 0, the conclusion is that 61 is either prime or pseudoprime, as anticipated.

At this point you may wish to try a calculation yourself. So, verify that

$$2^{10} \bmod 341 \;=\; 1,$$

then use this fact to show that

$$2^{340} \bmod 341 \;=\; 1.$$

This result tells you that the number 341 is either prime or pseudoprime. (In this case, as mentioned earlier, 341 is in fact a pseudoprime.)

The ARCL test works by altering the Fermat test so that it cannot be 'fooled' by a pseudoprime. It is this alteration that requires so much deep mathematics. (If you *really* want to see for yourself, the place to look is the article 'Primality testing and Jacobi sums', by Cohen and Lenstra, in the mathematical research journal *Mathematics of Computation*, Volume 42 (1984), pp. 297–330.)

Mersenne Primes

The ARCL test is the fastest general-purpose primality test currently available. The phrase 'general purpose' here means that it will work on any given number n. But for numbers having special structures there are often alternative methods which are much faster, the process being speeded up by exploiting the special structure of the number. The most spectacular example of this concerns numbers of the form $2^n - 1$. Such numbers are nowadays called *Mersenne numbers* after a seventeenth-century French monk, Marin Mersenne.

In the preface of his book *Cogitata Physica-Mathematica*, published in 1644, Mersenne stated that the number

$$M_n = 2^n - 1$$

is prime for $n = 2, 3, 5, 7, 13, 17, 19, 31, 67, 127, 257$, and composite for all other n less than 257. How did he do it? No one knows. But at any rate he was astonishingly close to the truth. Only in 1947, when desk-top calculators became available, was it finally possible to check his claim. He had made just five mistakes: M_{67} and M_{257} are not prime, and M_{61}, M_{89}, and M_{107} are prime.

Mersenne numbers provide an excellent method of obtaining very large prime numbers. The rapid growth of the function 2^n as n gets larger guarantees that the Mersenne numbers M_n soon become extremely large, so the idea is to look for values of n for which M_n is prime. Such primes are called *Mersenne primes*. A bit of elementary algebra indicates that M_n will not be prime unless n itself is prime, so it is necessary only to look at prime values of n. But even most primes n give rise to a composite Mersenne number M_n, so the search for suitable values of n is not easy – though this is not at all apparent from the first few cases, since

$$M_2 = 2^2 - 1 = 3,$$
$$M_3 = 2^3 - 1 = 7,$$
$$M_5 = 2^5 - 1 = 31,$$
$$M_7 = 2^7 - 1 = 127,$$

are all prime. But then the pattern breaks, with

$$M_{11} = 2047 = 23 \times 89.$$

Then follow three more prime values:

$$M_{13} = 8191, \quad M_{17} = 131\,071, \quad M_{19} = 524\,287.$$

After that, Mersenne primes become harder to find. The next five values of n for which M_n is prime are 31, 61, 89, 107, 127.

When they see the above values for the first time, most people jump to the conclusion that if p is itself a Mersenne prime, then M_p is also prime. It certainly works at first: 3 is a Mersenne prime and so is M_3; 7 is a Mersenne prime and so is M_7; 31 is a Mersenne prime and so is M_{31}; likewise for 127 and M_{127}. But there the pattern stops. Though 8191 is a Mersenne prime (being M_{13}), M_{8191} (which has 2466 digits) is composite. This was discovered in 1953 using an early computer. (See the section on perfect numbers later in this chapter.)

In fact there are to date only thirty known Mersenne primes. The twelve values of n listed above for which M_n is prime were all known by the early years of this century. The next five ($n = 521, 607, 1279, 2203, 2281$) were all found in 1952 by Raphael Robinson using the SWAC computer. The value $n = 3217$ was discovered in 1957 by Hans Riesel using a BESK computer. In 1961, Alexander Hurwitz used an IBM 7090 computer to obtain the values $n = 4253$ and 4423, and in 1963 Donald Gillies and the ILLIAC-II found $n = 9689, 9941$, and 11 213. Bryant Tuckerman's IBM 360–91 bagged $n = 19937$ in 1971. With the next discovery, in 1978, record prime numbers became front-page news with the announcement that after three years' work involving 350 hours of computer time on the CYBER 174 at the California State University at Hayward, two 18-year-old high-school students, Laura Nickel and Curt Noll, had found the 6533-digit Mersenne prime M_{21701}. One year later, Noll bettered the record with the 6987-digit prime M_{23209}. Later the same year the record fell again, this time to David Slowinski, a young programmer working for Cray Research at Chippewa Falls, Wisconsin. Using the immensely powerful CRAY–1 computer, he found the 13 395-digit prime M_{44497}. In 1982 the same man–machine combination showed that M_{86243} (a 25 962-digit number) is prime. Then, moving on to the even more powerful CRAY–XMP computer, Slowinski went even higher in 1983 with the 39 751-digit prime M_{132049}. Finally (for now), in September 1985, in Houston, Texas, a CRAY–XMP owned by Chevron Geosciences found the current record holder, the 65 050-digit number M_{216091}. (Since Chevron were running Slowinski's Prime Finder program, the credit for the discovery really goes to him. The company were actually running the program because it provides a good way of showing up any faults in the computer system.)

Is this the end of the story? Probably not. It is conjectured that there is no end to the Mersenne primes – that there are infinitely many of them. But this has not been proved, and all that is known *for certain* is that there are at least thirty of them (i.e. the ones so far identified).

The method used to check Mersenne numbers for primality is very simple (though the mathematics behind it is not). It is known as the *Lucas–Lehmer test* after Edouard Lucas (who discovered the basic idea in 1876) and

Derrick Lehmer (who refined the method in 1930). To test if the Mersenne number M_n is prime (assuming n is already known to be prime), calculate the numbers $U(0), U(1), \ldots, U(n-2)$ by the rules:

$$U(0) = 4,$$

$$U(k+1) = [U(k)^2 - 2] \bmod M_n.$$

If at the end you find that $U(n-2) = 0$, then M_n is prime. If $U(n-2) \neq 0$, then M_n is not prime.

For example, suppose we wanted to use the Lucas–Lehmer test to see if M_5 is prime. (Since $M_5 = 2^5 - 1 = 31$, we know already in this simple case that it is a prime, of course, but it will illustrate the method.) Then we make the calculation:

$$U(0) = 4,$$

$$U(1) = (4^2 - 2) \bmod 31 = 14 \bmod 31 = 14,$$

$$U(2) = (14^2 - 2) \bmod 31 = 194 \bmod 31 = 8,$$

$$U(3) = (8^2 - 2) \bmod 31 = 62 \bmod 31 = 0.$$

Since $U(3) = 0$, M_5 must be prime.

You might want to try this out for yourself using the two numbers $M_7 = 127$ (which is prime) and $M_{11} = 2047$ (which is not prime – see earlier).

Factoring

At the October 1903 meeting of the prestigious American Mathematical Society, the mathematician Frederick Nelson Cole was listed in the programme as presenting a paper with the rather unassuming title 'On the factorization of large numbers'. When called upon to speak, Cole walked up to the blackboard and, without saying a word, performed the calculation of 2 raised to the power of 67, following which he subtracted 1 from the result. Still saying nothing he moved to a clean part of the board and multiplied together the two numbers

$$193\,707\,721 \text{ and } 761\,838\,257\,287.$$

The answers to the two calculations were the same. Cole resumed his seat still having uttered not one word, and for the first and only time on record, the entire audience at an American Mathematical Society meeting rose and gave a 'speaker' a standing ovation.

What Cole had done (and apparently it took him twenty years of Sunday afternoons) was to find the prime factors of the Mersenne number M_{67}. It had been known since 1876 that M_{67} was composite. But this had been discovered (by Edouard Lucas himself) using the Lucas (now Lucas–Lehmer) test, which, although it provides an answer to the question of whether a given Mersenne number is prime or composite, gives no information about the factors of any number found to be composite. (The same is true of the ARCL test, as can be appreciated from the outline of this method given earlier, and indeed for any of a number of fast primality-testing methods currently available.)

Just how do you go about finding the factors of a number which you know is composite? Trial-and-error is clearly out of the question for the same reason that it is not practicable as a test for primality. But in practice there is an element of trial division in all current implementations of primality tests and factoring methods. Because it can be done quickly, it makes sense to start by trial division with, say, the first million prime numbers. If a divisor is found, then both the primality and the factorization problems are solved. If not, then at least you know that the number is either prime or else, if composite, has only large prime factors. Use of this last fact is made in a simple factoring method due to Fermat, described next.

Suppose that $n = uv$, where u and v are both large, odd numbers, say $u \leqslant v$. (As we are assuming that n has only large prime factors, this is the situation that will face us when we know that n is composite and want to find the factors.) Let

$$x = \tfrac{1}{2}(u + v), \qquad y = \tfrac{1}{2}(u - v).$$

Then $0 \leqslant y < x \leqslant n$, and $u = x + y$, $v = x - y$, so

$$n = (x + y)(x - y) = x^2 - y^2,$$

which can be rewritten as

$$y^2 = x^2 - n. \tag{1}$$

Conversely, if x and y satisfy Equation (1), then n has the factorization

$$n = (x + y)(x - y). \tag{2}$$

Hence, factoring n into a product of two numbers is equivalent to finding numbers x and y which satisfy Equation (1), in which case the resulting factorization is given by Equation (2). (Notice that this does not necessarily yield the *prime* factorization of n. But once a number has been split into two factors, they in turn may be factored, a task which is invariably much easier since the smaller a number, the easier it is to factor.)

To find x and y as in Equation (1), start with the smallest number k such that $k \geqslant \sqrt{n}$, and then try each of the values $x = k$, $x = k + 1$, $x = k + 2$, ... in turn, checking each time to see if $x^2 - n$ is a perfect square. Once such an x is found, the factorization is effectively completed, of course. Provided n has two factors of approximately the same size (hence close to $\sqrt{n}$ where the method starts off), a solution should be found fairly quickly.) (If you want to try the method yourself at this stage, the numbers 10 379 and 93 343 provide good examples.)

There are various ways of speeding up the above process. For instance, if you were doing it by hand, there is no need to try to evaluate the square root of $x^2 - n$ in every case to see if it turns out to be a whole number. Since no perfect square ends in any of the digits 2, 3, 7, or 8, whenever $x^2 - n$ is found to end with such a digit that value of x can be ignored at once.

Fermat himself used this method to obtain the factorization

$$2\,027\,651\,281 = 44\,021 \times 46\,061.$$

Computer implementations use some quite sophisticated methods for 'instantly eliminating' impossible values of x (a process known, for obvious reasons, as *sieving*). In 1974, mathematicians at the University of California at Berkeley built a specially designed electronic device for sieving numbers, the SRS–181, which can process 20 million numbers a second.

Fermat Numbers

The nth *Fermat number* is obtained by raising 2 to the power n, then raising 2 by that number and adding 1 to the result, i.e.

$$F_n = 2^{2^n} + 1.$$

Thus $F_0 = 3$, $F_1 = 5$, $F_2 = 17$, $F_3 = 257$, and (already the rapid growth of these numbers due to the repeated application of the exponential function is becoming apparent) $F_4 = 2^{16} + 1 = 65\,537$.

Interest in these numbers arose because of a claim made by Fermat in a letter written to Mersenne in 1640. Having noted that each of the numbers F_0 to F_4 is prime, Fermat wrote: 'I have found that numbers of the form $2^{2^n} + 1$ are always prime numbers and have long since signified to analysts the truth of this theorem.' This remark should serve as a warning to all who come to a conclusion on the basis of a small amount of information. For all his great abilities with numbers, Fermat was wrong in his claim. This was first shown conclusively by the great Swiss mathematician Leonhard Euler in 1732: $F_5 = 4\,294\,967\,297$ is not prime. Although Euler obtained this result by trial division, there is an extra irony here in that a straightforward calculation using Fermat's own test will demonstrate the non-primality of F_5. The test is, remember, that if p is prime then $3^{p-1} \bmod p = 1$, but for $p = F_5$ you get $3^{p-1} \bmod p = 3\,029\,026\,160$, so F_5 cannot be a prime.

Subsequent work has shown just *how* wrong Fermat was. It is now known that F_n is composite for all values of n from 5 to 16, as well as for various other values, and the current guess is that F_n is composite for *all* values of n greater than 4.

Fermat numbers provide another example of numbers whose special form makes it possible to test for their primality in an efficient manner – one popular method being *Proth's theorem*: the Fermat number F_n is prime if, and only if,

$$3^{(F_n-1)/2} \bmod F_n = -1.$$

This result provides a very efficient test for the primality of a Fermat number. (It is, as you may have guessed, closely related to the Fermat test discussed earlier.) But our interest here lies not in the testing for primality of Fermat numbers, but in the factorization of the ones known to be composite, for it is in this area that recent years have seen some significant developments, developments not without applications outside the realm of mathematics. (See the section on secret codes later in this chapter.)

The Fermat number F_5 was, as we have already noted, proved to be composite by Euler. He also calculated a prime factor, namely 641. In 1880 F_6 was shown by Landry to be composite, and once again a prime factor was found: 274 177. With F_7 the story was somewhat different. It was proved to be composite by Morehead and Western in 1905, but it was not until 1971 when Brillhart and Morrison (armed with an IBM 360–91 computer) found the factorization

$$F_7 = 2^{128} + 1$$

$$= 340\,282\,366\,920\,938\,463\,463\,374\,607\,431\,768\,211\,457$$

$$= 59\,649\,589\,127\,497\,217 \times 5\,704\,689\,200\,685\,129\,054\,721.$$

To do this they used a method suggested much earlier by Lehmer and Powers involving continued fractions. The computation took about an hour and a half. Improved versions of the *continued fraction method* (as it is now called) provide some of the best factoring methods currently available.

The same two who showed that F_7 was composite in 1905, Morehead and Western, also found in 1909 that F_8 was composite. It was only in 1981 that Brent and Pollard found the factorization. The computation took two hours on a Univac 1100/42 computer. The method, devised by Pollard himself, was at the time unusual in that, unlike most methods in mathematics, it did not guarantee to produce a result. All that could be concluded from the background mathematics was that if a certain computation was performed, then it was highly likely that a factorization of the number would result within a reasonable length of time, but there was a small possibility that this would not happen. So the method is not like trial division, where the chance of getting an answer within a billion years is small. There is still an element of chance involved, but what was clever about Pollard's method was that the odds were heavily stacked in the favour of the would-be factorer. In recent years there have been a number of so-called *Monte Carlo methods*, such as Pollard's factorization technique, which trade off the certainty of a result against a high probability of a result in much less time.

The two prime factors of F_8 (which has 78 digits) are

$$1\,238\,926\,361\,552\,897$$

and

$$93\,461\,639\,715\,357\,977\,769\,163\,558\,199\,606\,896\,584$$
$$051\,237\,541\,638\,188\,580\,280\,321.$$

At the time of writing, no one has been able to factorize F_9. Had he been alive today, now fast computers are available, perhaps the German mathematician Karl Friedrich Gauss would have been able to help. He certainly produced what must surely be the most astonishing result on Fermat numbers, linking them with a classical problem of Greek geometry. This result, as with many of Gauss's discoveries, merits a special introduction to one of the most amazing mathematical minds the world has ever known.

An Amazing Mathematical Mind

Karl Friedrich Gauss was born at Brunswick, now in West Germany, in 1777. His father, a bricklayer, hoped that his son would be able to assist him in his work – both as a labourer and to keep the accounts. The latter task was one to which the young Gauss seemed highly suited when, as a three-year-old child, he was able to correct his father's payroll calculations. Fortunately for the future of mathematics (to say nothing of physics and astronomy), the reigning duke soon came to hear of the young child prodigy and arranged for his formal education. At fifteen years of age, having progressed far beyond the abilities of his schoolteachers, Gauss went to Caroline College. Within three years the professors there had to admit he had left them way behind as well.

It was whilst he was a college student, in 1796, that Gauss made his remarkable observation concerning Greek geometry and Fermat numbers. The result appeared in the seventh and last section of his mammoth work *Disquisitiones Arithmeticae*, a book – in print to this day – which appeared in 1801 when Gauss was still only twenty-four, and which forms the basis of much of present-day number theory (that branch of mathematics which deals with properties of the natural numbers, of which the material in this chapter is but a small part).

One of the favourite problems of the Ancient Greek mathematicians was the construction of plane figures (circles, triangles, parallelograms, and so on) using only a ruler (unmarked, and thus suitable only for drawing straight lines) and compasses (used only to draw arcs of circles, not to transfer a length by moving them across the page). By the exercise of often considerable ingenuity it is possible to construct a great many geometrical figures using just these two rudimentary tools. (Until the mid 1960s such constructions formed a significant part of the mathematical schooling of pupils around the world.) The Greeks themselves knew how to construct regular n-sided polygons for $n = 3, 4, 5, 6, 8, 10, 12, 15, 16$. (A polygon is *regular* if all its sides have the same length and all its internal angles are equal.)

What the nineteen-year-old Gauss proved was that a regular polygon with n sides can be constructed using just ruler and compasses if, and only if, either $n = 2^k$ for some number k or else $n = 2^k p_1 p_2 \ldots p_r$ (for some k) where $p_1, p_2, \ldots, p_r$ are distinct Fermat primes. In particular, for any Fermat prime p you can construct a regular polygon with p sides. For the first Fermat prime, $F_0 = 3$, you get an equilateral triangle, which is easy to construct, and for the next one, $F_1 = 5$, you get a regular pentagon. Since $F_2 = 17$ is also a Fermat prime, Gauss's result shows that a regular 17-sided polygon may also be constructed using ruler and compasses. This was the first (and only) advance on the problem of constructing regular polygons since the time of the Greeks themselves, and Gauss was so proud of his discovery that he asked for a regular 17-sided polygon to be engraved on his tombstone. Though this request was never fulfilled, such a polygon is inscribed on the side of a monument to him erected in Brunswick, his birthplace.

Perfect Numbers

As was noticed by the Pythagoreans (the followers of the sixth-century BC mathematician Pythagoras), the number 6 has a rather special property. It is equal to the sum of its own divisors (other than itself):

$$6 = 1 + 2 + 3.$$

The next number after 6 with this property is 28. The only numbers which divide 28 are 1, 2, 4, 7, 14, and 28 itself, and

$$28 = 1 + 2 + 4 + 7 + 14.$$

Such numbers were named *perfect numbers* by the Pythagoreans.

In his first-century AD *Introductio Arithmeticae*, the Greek mathematician Nicomachus lists four known perfect numbers, the third (after 6 and 28) being 496, and the next 8128. Two conjectures followed from this evidence: that the *n*th perfect number contains *n* digits, and that the perfect numbers end alternately in 6 and 8. Both conjectures are false. For a start there are no perfect numbers with five digits. Moreover, though the fifth perfect number does end in a 6, being 33 550 336, so does the sixth, which is 8 589 869 056. (It is however true that any perfect number ends in either 6 or 8. This can be proved directly, and does not depend on knowledge of which numbers actually *are* perfect.)

In Book IX of his *Elements*, Euclid proved in about 350–300 BC that if $2^n - 1$ is prime, then the number $2^{n-1}(2^n - 1)$ is perfect. Two thousand years later, Euler showed that every even perfect number is of this type. Thus was established the close relationship between Mersenne primes and perfect numbers, which implies at once that there are at present exactly thirty known even perfect numbers. In fact there are no known odd perfect numbers, and it is conjectured that all perfect numbers are of necessity even. Though this has not been proved, some evidence has been collected in favour of the conjecture. It is known that any odd perfect number, if one were to exist, would have to be larger than 10^{100} and have at least 11 distinct prime factors. On the other hand, if history is any guide one should be careful about making conjectures about perfect numbers. In his 1811 book *Theory of Numbers*, Peter Barlow wrote, of the eighth perfect number, $2^{30}(2^{31} - 1)$, a 19-digit number discovered by Euler in 1772, '[It] is the greatest that ever will be discovered; for as they are merely curious, without being useful, it is not likely that any person will ever attempt to find one beyond it.'

Though he appears to have been right about perfect numbers being merely of curiosity value, Barlow clearly underestimated the fascination of curiosities, as the first section of this chapter illustrates only too well. And there is no doubt that perfect numbers are very curious. For instance, every (even) perfect number is *triangular*; which means that it can be represented by a number of balls arranged to form an equilateral triangle (which is equivalent to being of the form $\frac{1}{2}n(n + 1)$ for some number *n*). Another fact: if you take any perfect number other than 6 and add together all its digits, the number you get will be one more than a multiple of 9. Related to this is the result that the *digital root* of any perfect number is 1. (To obtain the digital root you add together all the digits of the number, then all the digits of *that* number, and so on until you end up with a single-digit number.)

And again, every perfect number is a sum of consecutive odd cubes. For instance:

$$28 = 1^3 + 3^3,$$
$$496 = 1^3 + 3^3 + 5^3 + 7^3.$$

One more: if n is perfect, then the sum of the reciprocals of all the divisors of n is always equal to 2. For instance, 6 has divisors 1, 2, 3, 6, and

$$\frac{1}{1} + \frac{1}{2} + \frac{1}{3} + \frac{1}{6} = 2.$$

In fact, so much effort has been spent searching for these 'curious' numbers that, despite Barlow's claim as to their uselessness, their computation has acquired something of the status of a benchmark for measuring computer power. For instance, take the Mersenne number M_{8191}, the first number to break the chain of Mersenne primes giving rise to Mersenne primes (see earlier). Using the Lucas–Lehmer test to demonstrate that this 2466-digit number is not prime (and hence does not yield a perfect number) took 100 hours when first done on the ILLIAC-I computer in 1953. Over the years the computation time has come down dramatically: 5·2 hours on an IBM 7090, 49 minutes on ILLIAC–II, 3·1 minutes on an IBM 360–91, and 10 seconds on a CRAY–1.

Secret Codes

In the autumn of 1982, at a scientific meeting in Winnipeg, Canada, two mathematicians and a computer engineer went out for a beer one evening. The two mathematicians soon got around to talking about how you can factorize large numbers, and the computational problems that can arise. Hearing them, the computer engineer mentioned that the design of the particular machine he worked on would enable one of the major problems they had met with to be overcome quite easily. And so it was that a chance encounter in a bar was to have significant repercussions in the field of data security. For the difficulty of factoring large numbers lies at the heart of one of the most secure forms of secret code. The story of just how an apparently useless and esoteric piece of pure mathematics came to be the basis of modern security systems is one of the most interesting mathematical tales of this century, and a dramatic warning to anyone who declares that a particular piece of scientific work is 'of no practical use'.

Some of the worst offenders when it comes to decrying the usefulness of their subject are mathematicians themselves. Writing in his excellent little book *A Mathematician's Apology*, the great British mathematician G. H. Hardy said: 'Real mathematics has no effects on war. No one has yet discovered any warlike purpose to be served by the theory of numbers or relativity, and it seems very unlikely that anyone will do so for many years' (Chapter 28). This was written in 1940. By 1945 the world had seen the horrific disproof of Hardy's claim about warlike uses for relativity in the form of the atomic bomb. As to his other example, number theory, this 'useless' subject now provides the security systems which are used to control (and perhaps one day even launch) the hundreds of nuclear missiles which have proliferated since that first bomb was dropped on Hiroshima. So much for predicting applications (or lack of them) of mathematical discoveries in the world at large. Hardy's own subject was, incidentally, number theory itself, and some of his own work has proved to be of real use, despite his personal claim – made in Chapter 29 of the same book – that 'I have never done anything "useful". No discovery of mine has made, or is likely to make, directly or indirectly, for good or ill, the least difference to the amenity of the world.'

There is, of course, nothing new in the idea of secret codes. Julius Caesar used them in order to ensure the security of the orders he sent to his generals during the Gallic wars. Nowadays it is not only the military who require their communications to be made secure by encryption techniques. There are also commercial and political reasons for ensuring that information does not pass into the wrong hands.

How do you go about designing an encryption system? A not entirely flippant answer would be 'With great care'. The would-be cryptanalyst (i.e. the 'enemy' who is trying to break your code) has a great many weapons at his disposal, in the form of both powerful computational equipment and sophisticated mathematical and statistical techniques. Certainly the simple kinds of cipher used by Caesar are woefully inadequate. In a *Caesar cipher* the original message is transformed by taking each letter of each word in turn and replacing it by some other letter according to some fixed rule, such as taking the letter three places along in the alphabet, so A is replaced by D, G by J, Y by B, and so on. Thus the word MATHEMATICS would become PDWKHPDWLFV. A message encrypted in this way may look on the surface to be totally indecipherable without knowing the rule used, but this is by no means the case. For one thing, there are only 25 such 'shift along' ciphers, and an enemy who suspected you were using one need only try them all in turn until the one used was found. But even if you employ some other, less obvious rule for substituting letters the resulting code will not be secure. The problem is that there are very definite frequencies with which the individual letters occur in English (or in any other language), and by counting the number of occurrences of each letter in your coded text an enemy can easily deduce just what your substitution rule is – especially when computers are used to speed up the process.

With simple substitution out of the question, what else might you try? Whatever you choose the same dangers are present. If there is any kind of 'recognizable' pattern to your coded text, a sophisticated statistical analysis can usually crack the code without much difficulty. And now the real difficulty becomes apparent. In order that your message can be successfully decoded at the other end (possibly thousands of miles away), the trans-formation performed on the message by your encryption scheme clearly must not destroy *all* order – the message must still be there beneath it all. And yet this hidden order has to be buried sufficiently deeply to prevent an enemy from discovering it.

All modern cipher systems use computers; they have to. The enemy may be assumed to have powerful computers to analyse your message, so your system needs to be sufficiently complex to resist computer attack. Because of the difficulty of designing and maintaining the security of cipher systems, they invariably consist of two components: an encryption procedure and a 'key'. The former is, typically, a computer program or possibly a specially designed computer. In order to encrypt a message the system requires not only the message but also the chosen key, usually a secretly chosen num-ber. The encryption program will code the message in a way which depends upon the chosen key, so that only by knowing that key will it be possible to decode the ciphered text produced (see Figure 2). Because the security depends on the key, the same encryption program may be used by many people for a long period of time, and this means that a great deal of time and effort can be put into its design. A helpful analogy is that manufacturers of

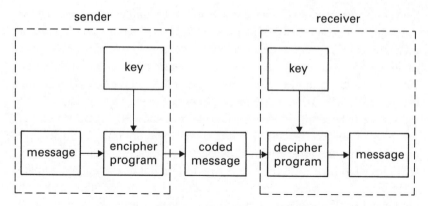

Figure 2. A typical cipher system. The encipher program (which may be in the form of specially made equipment, or a program for a general-purpose computer) uses a secret key chosen by the user in order to produce the encoded text. At the receiving end a similar system operates. Traditional systems employ the same key for both encryption and decryption, and the decipher program simply reverses the steps involved in the cipher program. Public key systems utilize two different keys, and the relationship between encipherment and decipherment depends on the mathematics involved.

safes and locks are able to stay in business by designing one type of lock which may be sold to hundreds of users, who rely upon the uniqueness of their own key to provide security. (The 'key' here could be of the combination variety, which would highlight at once the similarity between the two uses of the word 'key' in this discussion.) Just as an enemy may know how your lock is designed and yet still be unable to break into your safe without knowing the combination, so the enemy may know what encryption system you are using without being able to crack your coded messages – a task for which knowledge of your key is required.

In a typical 'key system' the message sender and receiver agree beforehand on some secret key which they then use to send each other messages. As long as they keep this key secret the system should be (if it is well designed) secure. One such system is the American-designed Data Encryption Standard (DES), which requires for its key a number whose binary representation has 56 bits (in other words, a string of 56 zeros and ones operates as the key). Why such a long key? Well, no one makes any secret of how the DES system works. All the details have been published. So in theory an enemy could crack your coded messages simply by trying all possible keys one after the other until one is found which works. With the

DES, there are 2^{56} possible keys to be tried, a number which is so large as to render the task virtually impossible. (Actually this figure is not quite high enough to provide cast-iron security, but with any cipher system a trade-off has to be made between security and convenience for the user. The bigger the key, the more cumbersome the process becomes.)

Though the DES enjoys widespread use at the moment, such systems have an obvious drawback. Before it can be used, the sender and receiver have to agree on the key they will use, and since they will not want to transmit that key over any communication channel, they have to meet and choose the key, or at the very least employ a trusted courier to convey the key from one to the other. Such a system is no good for communication between individuals who have not already met. In particular, it is not suitable for use in, say, international banking or commerce, where it is often necessary to send secure messages across the world to someone the sender has never met.

In 1975, Whitfield Diffie and Martin Hellman proposed a quite new type of cipher system: public key cryptography, in which the encryption method requires not one but two keys – one for enciphering and the other for deciphering (like having a lock which requires one key to lock it and another to unlock it). Such a system would be used as follows. A new user would purchase the standard program (or special computer) used by all members of the communication network concerned. He would then generate two keys. One of these, his deciphering key, he would keep secret. The other key, the one used for encoding messages *sent to him* by anyone else in the network, he would publish in a directory of the network users. To send a message to a network user, all that has to be done is to look up that user's public encipherment key, encrypt the message using that key, and send it. To decode the message it is of no help knowing (as everyone will) the enciphering key. You need the deciphering key. And only the intended receiver knows that. (So even the message sender cannot decode the message once it has been enciphered!)

Well, this all sounds fine, but how is such a system to be constructed? On the face of it, it seems impossible. The key (if I may use the word here) is to exploit the strengths and weaknesses of those very computers whose existence makes the encryptioner's task so difficult. As has been indicated earlier in this chapter, finding large primes (say of the order of 50 digits) is relatively easy. So is the task of multiplying two such large primes together to produce a single (composite) number (of around 100 digits or more). But factoring that number into its two primes is not at all easy, and indeed to all intents and purposes is impossible. This is the idea behind the most common implementation of a public key system in use today – the one designed by Ronald Rivest, Adi Shamir, and Leonard Adleman of the Massachusetts Institute of Technology and known by their initials as the RSA system. The secret deciphering key consists (essentially) of two large prime numbers chosen by the user (chosen with the aid of a computer – not chosen from any published list of primes, which an enemy might have access to!). The public encipherment key is the product of these two primes. As there is no quick method of factoring large numbers, it is practically impossible to recover the deciphering key from the public encipherment key. Message encryption corresponds to multiplication of two large primes (easy), decryption to the opposite process of factoring (hard). (This is not quite how the system works. Some mathematics is required which, though moderately sophisticated, was all known to Fermat. The important point is that since deciphering a message is just the opposite of enciphering it, the same has to be true of the relationship between the two keys.)

This is why the security of large international data networks nowadays relies upon the inability of mathematicians to find an efficient method of factoring large numbers (whilst at the same time being able to produce large primes with ease). Obviously the security of such a system depends on (amongst other things not mentioned here) the continued difficulty of factorization. And that was where the Winnipeg beer came in. The designers of the system originally suggested that two primes of around 50 digits each should provide sufficient security. (As always with such systems, the bigger the numbers used, the more expensive they become to run, so a good compromise value is sought.) Up to 1982, the best factoring methods developed were able to handle numbers of around 50 digits (using computers like the CRAY–1). By spotting how the particular design of the CRAY–1 arithmetic units could be exploited to overcome one of the problems they were up against, computer engineer Tony Warnock gave factoring experts Marvin Wunderlich and Gus Simmons just the information they needed in order to extend their models to handle numbers of 60 to 70 digits. Suddenly the RSA systems looked a little less secure. Though the answer was obvious – simply use primes of 100 digits each to produce a 200-digit public key – this unexpected advance created a ripple of uncertainty in the communications security business. This feeling of 'insecurity' has not been helped by other recent advances in factorization, for although 90-digit numbers appear to be the current upper limit of what is (*in general*) possible in factorization, the amount of sophisticated mathematics which is currently being brought to bear on the problem could at any time bring a real breakthrough.

Suggested Further Reading

Excellent coverage of number theory (though not of computational aspects) is provided by *Elementary Number Theory*, by David Burton (Allyn and Bacon, 1980).

Cipher Systems, by Henry Beker and Fred Piper (Northwood, 1982) covers most aspects of cryptography, and though hard going in places is well worth looking at.

2 Sets, Infinity, and the Undecidable

New Horizons

Sometimes the solution of a long-standing problem marks the end (or the beginning of the end) of a mathematical era or field, the culmination of years of effort. On other occasions it may open up an entire new area of research, possibly previously undreamt of. Such was the case with the discovery in 1963, by the 29-year-old Stanford University mathematician Paul Cohen, of the solution to *Cantor's continuum problem*. Not only was the nature of the solution itself of a revolutionary new kind, but also the methods Cohen developed in order to obtain it were quite new. These methods were soon found to have a very wide range of possible applications, and the ensuing twenty-year period saw tremendous activity based upon Cohen's breakthrough. In recognition of his work, Paul Cohen was awarded in 1966 a Fields Medal, the highest prize that can be bestowed on a mathematician, and equivalent to a Nobel Prize in the other sciences.

Prior to 1963, a mathematician faced with trying to determine the truth or falsity of some mathematical statement had two possibilities open to him: either prove it true or prove it false. Experience and intuition are often the only guide as to which of these is worth the greater effort, and a bad choice can lead to an immense amount of time being spent trying to achieve the impossible. But it was always felt that at the end of the day an answer would be reached. What Cohen did was destroy for ever this comforting belief, by demonstrating that there are mathematical statements which are neither true nor false, but *undecidable*. (Actually this is not quite accurate. The existence of undecidable statements had been established by Kurt Gödel in 1930 but, as will be explained later, this was not thought to affect ordinary, everyday, run-of-the-mill mathematics in the way that Cohen's result did.)

In order to explain just what it was that happened in 1963, it is necessary to go right back to the very nature of mathematics itself, and to a pioneering concept from the turn of the century.

The Axiomatic Method

Faced with determining the truth or falsity of some statement, the physicist, chemist, or biologist – indeed almost any scientist – will usually set up some experiment, or at the very least will make use of some reasoning which depends upon experimental evidence. It has to be so. Each of these scientists is studying some aspect of the physical world, and consequently it is that physical world which is the final arbiter of what is and what is not. But what happens in mathematics?

At the most basic level, mathematics is much like any other physical science in that certain aspects of the world around us are singled out for detailed study. So the world itself can provide some information. If you want to check that the sum of the angles of a triangle is 180°, you can go out and measure the angles of lots and lots of triangles. Though this would clearly provide a verification of the kind accepted by the physicist or chemist, it is, however, not how the mathematician proceeds in practice; nor would such a procedure constitute a *mathematical* verification of the assertion that the sum of the angles of *any* triangle is 180°.

The reason why a 'crude', experimental approach is not adequate for determining mathematical truth lies in the nature of what mathematics is and is intended to be. Though its roots lie in the physical world, mathematics is a precise and idealized discipline. The 'points', 'lines', 'planes', and other ideal constructs of mathematics have no exact counterpart in reality. (Chapter 4 sheds some interesting light on this point.) What the mathematician does is to take a totally abstract, idealized view of the world, and reason with his abstractions in an entirely precise and rigorous fashion. A simple (but important) example should help to make this clear.

One of the most basic mathematical abstractions is that of number. It is via numbers that most of us make our first encounter (usually at a very early age) with mathematical abstractions. By a process which seems nigh on miraculous when you come to think about it, as small children we all come to recognize that there is something in common between a collection of three apples, a collection of three uncles, a collection of three flowers, and so on. This abstraction of 'threeness' leads to the formation of the mental concept of the number three. What makes this seem miraculous is that there is, of course (!), nothing in the world which *is* the number three. It is purely an abstract concept, and it is only our familiarity with it that results in our feeling no unease (or even embarrassment) when we speak of 'the number three' – or any other number. (And you realize just how abstract the concept is when you try to explain what the number three is without using the word 'three' or 'threeness'. It cannot be done, yet none of us worries about that.) The same is true of all the other mathematical abstractions: though they may have their roots in the real world, the abstractions themselves are purely concepts, having no existence outside our minds.

Already with the concept of number we have the basis of the mathematical discipline of *number theory* discussed in the previous chapter. But how do you begin to handle abstractions at all, let alone in a manner which will prove to be of use in the physical world? The only way is to begin by laying down some ground rules. For numbers this amounts to formulating *postulates* (or *axioms*) which will be assumed to hold for all numbers, and then proceeding to make deductions from these initial postulates by means of rigorous, logical reasoning alone. (It is also possible to write down postulates governing the logical reasoning process used. This is the business of the branch of mathematics known as mathematical logic, which is closely connected with the subject-matter of this chapter.) For example, we know from experience that when two numbers are added together, the order in which they are taken is immaterial. For example, $5 + 3$ is the same as $3 + 5$. So a reasonable axiom to include in a system for arithmetic is the statement:

$$\text{for any numbers } m, n, \quad m + n = n + m.$$

This particular statement is known as the *commutative law of addition*. Another example is the *associative law of addition*:

$$\text{for any numbers } m, n, k, \quad (m + n) + k = m + (n + k).$$

Again, this arises from observations of the way addition works in practice. For example, if you want to add together the three numbers 3, 5, and 10, it does not matter if you first add 3 to 5 (to get 8) and then add 10 to the result, or else begin by adding 5 to 10 (to get 15) and then add 3 to that result. In either case you get the same answer, 18.

In adopting the above two axioms we are already taking a big step, a step which really amounts to nothing other than an act of faith. Though both axioms can be tested ('experimentally') by examining a large number of cases, there is not even the theoretical possibility of checking every case since there are infinitely many of them. Can we therefore be sure that the two axioms remain true when the numbers involved are very large, say involving millions of digits? It seems reasonable, possibly even 'obvious'. But then mathematics (and most other disciplines) is full of examples of 'obvious truths' which turn out to be false. (On the basis of common experience it does, after all, seem 'obvious' that the Sun orbits the Earth.) The available evidence can only *suggest* that the two axioms are true. There is no possibility of ever being able to *prove* this conclusively; their truth has to be *assumed*. That is why such assumptions are referred to as axioms – from the Latin word 'axioma', meaning a principle. (There is a sense in which both the above axioms can be proved, in that it is possible to write down 'more fundamental' postulates for the natural numbers and deduce the more familiar rules of arithmetic from them, but this only shifts the act of faith one step back, it does not eliminate it.)

To emphasize the point being made in the above paragraph, it is perhaps worth mentioning that although both laws considered are accepted axioms for the arithmetic of the integers, the commutative law is false for certain systems of infinite numbers (see later in this chapter) and the associative law fails when applied to computer arithmetic. (The law fails when very large numbers are added to very small ones.)

At this stage it is perhaps worth while to look at the axiomatic development of a mathematical theory in a little more detail. Since we have already examined some aspects of it, the arithmetic of the integers (i.e. positive and negative whole numbers) will provide an excellent example.

The Integers: An Example

The following axioms are adequate for the study of the basic arithmetic (i.e. addition and multiplication) of the integers.

(1) For all m, n, $m + n = n + m$ and $mn = nm$. (Commutative laws of addition and multiplication.)

(2) For all m, n, k, $(m + n) + k = m + (n + k)$ and $(mn)k = m(nk)$. (Associative laws of addition and multiplication.)

(3) For all m, n, k, $m(n + k) = (mn) + (mk)$. (Distributive law.)

(4) There is a number 0 which has the property that for any number n, $n + 0 = n$. (Existence of an additive identity.)

(5) There is a number 1 which has the property that for any number n, $n \times 1 = n$. (Existence of a multiplicative identity.)

(6) **For every number n there is another number k such that $n + k = 0$.** (Existence of additive inverses.)

(7) For any m, n, k, if $k \neq 0$ and $km = kn$, then $m = n$. (Cancellation law.)

Starting from these axioms, it is possible to prove all the usual properties of arithmetic of the integers. For instance, there is a rule analogous to Axiom 7 which applies to addition:

$$\text{if } k + m = k + n, \text{ then } m = n.$$

To prove this, start with the assumption that $k + m = k + n$. Then, by Axiom 1,

$$m + k = n + k.$$

By Axiom 6, let l be a number such that $k + l = 0$. Then, by adding l to both sides of the above equation, we get

$$(m + k) + l = (n + k) + l.$$

So, by Axiom 2,

$$m + (k + l) = n + (k + l).$$

In other words, when the choice of l is taken account of,

$$m + 0 = n + 0.$$

Using Axiom 4, it follows at once from this last equation that

$$m = n,$$

as required.

Again, to prove the result that $x \times 0 = 0$ for any number x, we argue as follows.

$$x + 0 = x \quad \text{(by Axiom 4, with } n = x\text{)},$$
$$= x \times 1 \quad \text{(by Axiom 5, with } n = x\text{)},$$
$$= x \times (1 + 0) \quad \text{(by Axiom 4, with } n = 1\text{)},$$
$$= (x \times 1) + (x \times 0) \quad \text{(by Axiom 3, with } m = x, n = 1, k = 0\text{)},$$
$$= x + (x \times 0) \quad \text{(by Axiom 5, with } n = x\text{)}.$$

So, by the additive analogue of Axiom 7 just proved (with $k = x$, $m = 0$, $n = x \times 0$),

$$0 = x \times 0.$$

At this stage you might like to try your hand at proving each of the following standard facts about integer arithmetic. In each case you should make sure that you use only facts already known, either because they are axioms or because you have proved them.

(1) There is exactly one element 0 which satisfies the requirements of Axiom 4: i.e. if $\varnothing$ has the property that $n + \varnothing = n$ for every number n, then $\varnothing = 0$. (Uniqueness of 0.)
(2) There is exactly one element 1 which satisfies the requirements of Axiom 5. (Uniqueness of 1.)
(3) For every pair m, n, there is exactly one number k such that $n + k = m$.

Notice that the last result above guarantees that subtraction is always possible in the integers (since the unique number k will be $m - n$), even though no mention was made of this operation in the axioms themselves. A particular case of the result is when $m = 0$, proving the uniqueness of k in Axiom 6.

Of course, if it were necessary to prove everything in mathematics in the same detail as above, the mathematician's task would be practically impossible. What makes it all work is the fact that mathematical knowledge is cumulative: once something has been established it may thereafter be used without further ado. (A very simple example of this phenomenon occurred in the above proof that $x \times 0 = 0$.) Consequently it is only at the very beginning of an axiomatically based development that such detailed proofs are necessary. For the most part mathematical reasoning is much more like a rigorous version of the 'everyday logic' employed in any other science.

Consistency, Completeness, and Truth

The bulk of present-day mathematics consists of making deductions from axioms. These axioms do not have to relate (directly, or even indirectly) to anything in the physical world. The axioms for integer arithmetic given in the previous section were obtained by examining the behaviour of the operations of addition and multiplication on those integers that are familiar to us. (This really means the small integers, though it should be remembered that it was only in the eighteenth century that *negative* numbers became widely accepted – see Chapter 3.) But once the axioms have been decided on, all questions as to their 'universal truth' become irrelevant, as do questions about just what *are* the objects the axioms refer to. For instance, nowhere in the axioms given in the previous section is any mention made of what exactly *is* a number. In fact there are many other collections of mathematical objects which also turn out (as a consequence of their formal definition) to satisfy these axioms. Because axiom systems often apply in many different situations, mathematicians frequently introduce names to describe the structures satisfying a particular axiom system. Any mathematical structure which satisfies the axioms of the previous section is said to be an *integral domain* (if Axiom 7 is omitted it is called a *ring*). Thus, in order to stipulate that the integers (together with their arithmetical operations of addition and multiplication) satisfy these axioms, it would be enough to say that they constitute an integral domain. The rational numbers (fractions), the real numbers, and the complex numbers provide other examples of integral domains.

Any result proved by means of logical argument starting from a given axiom system will be 'true' for the abstract structures that satisfy that axiom system, but questions about the 'truth' of the result in the real world would not just be unanswerable, *they would have no meaning*. If the axioms provide a good reflection of some phenomenon in the real world, then the consequences of those axioms will also relate well to the real world, and may even provide some useful information which can benefit (or possibly lead to the destruction of) the human race. But as far as the mathematics is concerned, the relevance or otherwise of the initial axioms is immaterial. Some axiom systems which have led to very interesting mathematics appear to have no relation whatsoever to the physical world – though this is not to say that a connection will not one day be discovered! At the price of isolating himself from reality (to a greater or lesser extent), the mathematician is able to work in a world of absolute certainty, with the potential bonus of his results finding widespread application (within mathematics itself in the first instance) because his axiom system fits structures other than the one he (may have) had in mind.

But if 'truth' cannot be the guide, what considerations do govern the formulation of an axiom system?

An essential requirement is *consistency*: it must not be possible to deduce two mutually contradictory consequences from the axioms. This requirement has to be satisfied by all axiom systems, though proving consistency is not only difficult, but fraught with philosophical problems as well (as will be indicated presently).

The other requirement, which is pertinent to any axiom system which attempts to describe some particular mathematical structure (such as the arithmetic of the integers), is *completeness*. The axiom system should be rich enough to enable all the 'true facts' about the structure to be provable.

Satisfying both of the above requirements involves a delicate balancing act. To achieve completeness, more and more axioms may need to be introduced. And yet the more axioms there are, the greater will be the likelihood of introducing an inconsistency.

The Gödel Incompleteness Theorems

At the turn of the century, the world-famous German mathematician David Hilbert proposed a programme for the development of all of mathematics within the strict formalization of the axiomatic method. According to Hilbert's belief, all of mathematics could be regarded as the formal, logical manipulation of symbols based on prescribed axioms. (This would mean that, in principle, a computer could be programmed to 'do all of mathematics'.) But in 1930, with two startling and totally unexpected theorems, the young Austrian mathematician Kurt Gödel demonstrated that Hilbert's programme could not possibly succeed.

What Gödel proved was that, for any consistent axiom system which is strong enough to allow for the development of elementary integer arithmetic, there will always be statements relevant to that axiom system which can be neither proved nor disproved from those axioms (First Incompleteness Theorem). Moreover, amongst those unprovable (from the axioms) statements is the statement that the axiom system is consistent – so the all-important (for Hilbert's programme) notion of consistency is destined to remain for ever elusive (Second Incompleteness Theorem).

Although Gödel's results meant that the axiomatic method could not be elevated to the all-embracing position envisaged by Hilbert, it should not be thought that it heralded the death of the axiomatic approach within every-day mathematics as it was, and still is, practised. On the contrary, this century has seen the axiomatic method assume a supreme position in mathematics. What Gödel forced us to abandon was the belief or hope that an axiom system could ever be adequate to answer all the questions we might reasonably ask of it.

In point of fact, with the increasing success of the axiomatic approach, the years after Gödel's pronouncement saw a gradual growth in the belief that it was only highly specialized statements which could not be proved. For instance, Gödel's first incompleteness theorem was obtained by show-ing that in elementary number theory it is possible to formulate a statement analogous to the – in English – obviously paradoxical statement:

> *The sentence enclosed in a box on this page is false.*

(In Gödel's number-theoretic analogue, 'false' is replaced by 'not provable'. It is because elementary arithmetic is necessary in order to formulate such a statement that Gödel's result applies only to axiom systems which provide this facility. But of course, any axiom system destined to fulfil the aims of the Hilbert programme would certainly have to enable you to perform elemen-tary arithmetic!)

Also, turning to Gödel's second incompleteness theorem, although the consistency of an axiom system is an important consideration, the fact that it cannot be proved from the axioms is not such a great problem. In writing down the axioms in the first place there is a tacit assumption of their consistency, and the main interest lies in the consequences of the axioms of a more 'solid' nature. For instance, in number theory the statement that the axioms are consistent is not of the kind that number theorists are generally concerned about. So perhaps Gödel's incompleteness results are not really so relevant after all. Or so it once seemed, before a brash young American came along and proved otherwise. Paul Cohen's devastating 1963 result completely shattered the cosy feeling that incompleteness did not affect 'real' problems, and it did so in the most basic and fundamental part of mathematics: *set theory*.

Axiomatic Set Theory

At the turn of the century the development of abstract, pure mathematics (and in particular the various subjects stemming from Newton's and Leibniz's *infinitesimal calculus*) that had taken place during the nineteenth century led the German mathematician Georg Ferdinand Ludwig Philipp Cantor to formulate a very general mathematical 'framework' which could serve as a foundation for all of mathematics. That subject created by Cantor is still with us, and serving as a good foundation. It is known as *set theory*, and its concepts and methods pervade practically all of present-day mathematics. But its development since Cantor's original formulation has been both violent and traumatic, as the following pages will reveal. Before that, however, it is necessary to say a little about formal logic.

At the same time as Cantor was developing his ideas of set theory, and in particular a system of infinite numbers to measure the 'size' of infinite sets, Gottlob Frege was creating what is now known as *predicate logic*. Broadly speaking, this provides a universal, formalized language which is adequate for the expression of any mathematical concept whatsoever. Not that the importance of this development stemmed from any great need or desire on the part of mathematicians to carry out their work using predicate logic. Indeed, owing to the simplicity of the language, in most cases the expression of a mathematical concept or argument within Frege's framework will be extremely long and cumbersome. Frege's work was important firstly because it demonstrated quite clearly that all the many branches of mathematics are part of a single, coherent whole, and secondly (and far more importantly) as it enabled a proper analysis to be made of the deductive methods used by mathematicians in constructing proofs. (Though it ought to be noted that recent times have seen increasing use being made of predicate logic for the actual expression of mathematical concepts and proofs, in attempts to develop computer programs to obtain, or help to obtain, mathematical results. Obviously, in order to present mathematics in a form suitable for a computer to handle, a precise and quite simple language has to be used, and predicate logic provides just such a framework.)

The concept of a set as used by Cantor was extremely simple. A *set* is any collection of objects, or at least any collection of mathematical objects. The key is then to regard the collection as *a single object in its own right*. Small, finite sets can be described by listing their *members* (or *elements*), usually enclosing the list between curly brackets. Thus

$$\{1, 3, 5, 9\}$$

denotes the set whose elements are the numbers 1, 3, 5, and 9. For larger (and possibly infinite) sets it is not possible to list all the elements, and then you have to rely on some *property* to determine what set you are thinking of. The standard notation for the set of all those objects x for which the property $P(x)$ is valid is

$$\{x \mid P(x)\}.$$

Thus, the set of all prime numbers (an infinite set) may be denoted by

$$\{x \mid x \text{ is a prime number}\}.$$

There will also be sets for which there is no defining property, and for such sets there can be no specification in terms of their elements, but this point is not really relevant to the present elementary discussion. Roughly speaking, such sets arise 'by default', since the notion of a collection does not entail the existence of a property which determines it – but this is an advanced, subtle point. Ignoring these elusive sets for now, in favour of those determined by properties, what kind of 'properties' are to be allowed in the formation of sets? The original answer was, as you might expect by now, any property which can be expressed in Frege's predicate logic, a definition which by the very nature of this formalized language is both precise and adequate for all the properties encountered in mathematics.

At this stage things could hardly seem more rosy. Set theory provides an adequate foundational framework upon which all mathematical objects and structures may be constructed, and Frege's predicate logic provides a universal language for defining and discussing these objects and structures, including the basic notion of a set itself. Frege himself made extensive use of set-theoretic concepts in his two-volume work *The Foundations of Arithmetic*, intended as the culmination of his entire life's work.

It was whilst the second volume of Frege's book was at the printers that he received a letter dated 16 June 1902 from the famous British logician Bertrand Russell. After an initial paragraph praising Frege's first volume, Russell came to the crux of his letter. 'There is just one point where I have encountered a difficulty', he begins. What comes next is a brief explanation of an observation he had made exactly one year earlier – an observation that completely destroyed Frege's entire theory.

Russell's paradox, as it is known, is as simple as it is profound. According to the basic principle of Cantor's set theory, if $P(x)$ is any property (express-ible in predicate logic) applicable to the mathematical object x, then there is a corresponding set of all those x for which $P(x)$ is true, i.e. the set

$$\{x \mid P(x)\}.$$

There is nothing to stop the objects x involved here being themselves sets, for a set is a mathematical object just like any other. (Indeed, when set theory is taken as the basic foundation of mathematics, every mathematical object turns out to be a set of one kind or another.) Russell now took for the property P the statement (applicable to sets x)

$$R(x): x \text{ is not a member of } x.$$

(The standard symbol for set membership is $\in$, so $x \in y$ means that x is a member of y, and non-membership is denoted by $x \notin y$, so Russell's property $R(x)$ may be written as $x \notin x$.)

Now let us give the set determined by property $R(x)$ a name, say y. Thus

$$y = \{x \mid x \notin x\}.$$

Since y is a set, it is entirely reasonable to ask whether y is a member of itself. If it is, then y must satisfy its own defining property, which is to say that $y \notin y$ – i.e. y *is not* a member of itself. On the other hand, if y is not a member of itself, then y cannot satisfy its defining property, so $y \in y$ must hold – y *is* a member of itself. So we have arrived at the obviously untenable situation where, if y is a member of itself then it is not, and if it is not a member of itself then it is. A genuine paradox.

What made the Russell paradox so devastating was its utter simplicity. It employed only the most basic concepts, concepts upon which practically all of mathematics depends.

A way out of the dilemma caused by Russell's paradox was provided by the German mathematician Ernst Zermelo, whose work on integral equations (a highly applicable area of mathematics) had led him to consider deep problems concerning the nature of infinite sets. In 1908, in order to establish a secure set-theoretical framework for his work, he published a paper in which he developed a system of axioms for set theory. Subsequently modified by Abraham Fraenkel, *Zermelo–Fraenkel set theory* gradually came to be accepted as the 'correct' axiomatic approach to the theory of abstract sets. (An adequate motivation and explanation of the axioms involved requires more space than is available here. An elementary account can be found in various texts – see later for details.)

By virtue of Gödel's incompleteness theorems, there is, of course, no possibility of *proving* that the axioms of Zermelo–Fraenkel set theory are consistent, but they certainly appear to avoid paradoxes such as Russell's paradox, and most mathematicians believe that they will not lead to any contradiction at all – a belief which has grown stronger as the theory has shown that it can stand the test of time and heavy usage.

So much for consistency. What about completeness? The incompleteness theorems also tell us that there will be statements about sets which can be neither proved nor disproved on the basis of the axioms adopted. This deficiency assumes a greater importance than usual by virtue of the special, foundational nature of set theory. Since the entire edifice of modern mathematics can be regarded as (and to a great extent explicitly *is*) built upon set theory, deficiencies within set theory might result in genuine deficiencies in other areas of mathematics. But though this was always a possibility, the Zermelo–Fraenkel axioms did appear to be adequate to provide a theory of sets sufficient for mathematics, and most working mathematicians quietly ignored this danger, assuming that it would not affect them. Until, that is, Cohen forced the issue out into the open with his breakthrough in 1963.

Though Cohen's discovery turned out to have many ramifications, it was initially concerned with a problem involving Cantor's infinite numbers, the theory of which became perfectly sound (as far as anyone knows!) once the Zermelo–Fraenkel axioms had been formulated. So the time has now come to journey into the infinite and take a look at Cantor's theory.

Infinite Sets

Despite the fact that the world we live in is finite, the mathematics we need in order to deal with it involves the infinite at almost every turn: the set of all natural numbers is an infinite set, the precise specification of the number π requires infinitely many decimal places, the number of points on the smallest of lines is infinite, and so on. Though attempts have been made to avoid all use of the infinite, the mathematics that results turns out to be incredibly cumbersome and unwieldy. For despite its total abstraction, the infinite is a world of great simplicity. Going from the finite to the infinite is very much like stepping back from a television screen: when you are far enough away, the indecipherable complexity of the large number of tiny light spots occupying the screen is recognized as a coherent picture. By going to the infinite, the complexity of the large finite is lost. This is a phenomenon not restricted to pure mathematics alone. In economics, for example, idealized economies with infinitely many traders are studied in preference to the very large finite economies of the real world, and in physics infinite volumes are used in discussing certain subtle concepts of heat and electrical energy.

It was the development of a system of infinite numbers and their arithmetic which formed the crowning achievement of Cantor's pioneering work on set theory. 'But why do we need infinite numbers at all?' you might ask. The answer is: for the same reason that we need finite (whole) numbers – to count the 'number' of members of a set. The natural numbers enable us to 'measure the size' of a finite set. In order to measure the size of an infinite set, infinite numbers are required. (From which you can anticipate that it is not enough simply to refer to such a set as 'infinite'.) Having accepted that point, you might then ask, what is an infinite number? To which a good response would be 'What *is* a finite number?' As we saw at the beginning of this chapter, the natural numbers are mere figments of our imagination, so postulating the existence of infinite numbers should be no different. What is important is the way these infinite numbers behave, and this is where the key behind Cantor's infinite numbers lies.

The natural numbers are abstracted from finite sets (either mathematical ones or real-life sets like sets of apples, sets of people, and so on). The number three is that which all sets of three elements have in common. On the face of it this looks like a circular definition (which would thus not be a definition at all, of course), but Cantor observed that this was not at all the case. Rather, before we define 'number', we must have the notion of 'same size' for two sets, as will now be explained.

Two sets, call them A and B, have the *same size* if it is possible to match their elements in such a way that each member of A is matched to exactly one member of B, and vice versa. Thus, for example, the sets

$$A = \{1, 2, 3, 4\}, \qquad B = \{100, \pi, \sqrt{2}, \tfrac{1}{2}\}$$

are the same size, as is testified by the matching (there are others):

$$
\begin{array}{cccc}
1 & 2 & 3 & 4 \\
\updownarrow & \updownarrow & \updownarrow & \updownarrow \\
100 & \pi & \sqrt{2} & \tfrac{1}{2}
\end{array}
$$

Similarly, the sets

$$A = \{a, b, c\}, \qquad B = \{\text{foot, sock, shoe}\}$$

are the same size by virtue of the matching

$$
\begin{array}{ccc}
a & b & c \\
\updownarrow & \updownarrow & \updownarrow \\
\text{foot} & \text{sock} & \text{shoe}
\end{array}
$$

Notice that in neither case does the concept of the number of elements in the set arise. In order to talk about 'same size' it is not necessary to have a prior notion of 'size', nor is there any need to consider only finite sets. The same ideas apply to infinite sets (though in this case it will not be possible to describe the matching explicitly, as above). When applied to infinite sets, however, you soon find yourself faced with some unexpected results. For example, let A be the set of all natural numbers, and let B be the set of all even natural numbers. Intuitively, B is exactly 'half the size' of A, but

according to our definition these two sets are the same size, as testified by
the matching

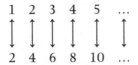

There is, however, no contradiction here – or if there is it is only with our
preconceptions. It is just that infinite sets do not always behave in the same
way as do finite sets.

A nice illustration of the kind of behaviour that arises with infinite sets
is provided by *Hilbert's Hotel*. This idealized institution has an infinite number of rooms, numbered 1, 2, 3, and so on all the way through the natural
numbers. One night, as chance would have it, all the rooms are taken. (In this
story there are infinitely many people as well.) And yet a latecomer can still
be accommodated, without anyone having to be thrown out. All that needs
to be done is to put the newcomer into Room 1, moving the present occupant of that room into Room 2, that room's occupant into Room 3, and so
on. All the guests already there are moved one room along, allowing the
latecomer to move into the now empty Room 1. (In fact it is possible to
accommodate infinitely many latecomers. Can you see how?) Though the
idea of an infinite hotel might seem far-fetched, there is nothing wrong with
the internal logic of the discussion. However counter-intuitive, that is the
kind of thing that happens when you start to explore the world of the
infinite.

The example of the natural numbers and the even natural numbers
might lead you to suspect that all infinite sets are the same size, which
would mean that there is no need for a system of infinite numbers. Indeed,
a great many of the infinite sets commonly encountered in mathematics are
the same size. For instance, the set of prime numbers, the set of natural
numbers, the set of integers, and the set of rational numbers all have the
same size. (Sets having the same size as the natural numbers are often
referred to as *countable*, since matching them with the natural numbers
provides a way of counting out their elements.) But, as Cantor discovered,
not all infinite sets are the same size. In fact there is a whole (infinite)
hierarchy of infinities, getting larger and larger all the while. Cantor's proof
of this key fact is both simple and elegant, using only the most basic notions
of set theory, but it is highly abstract, and for this reason will be left until
the very end of the chapter where the more squeamish readers may leave
it alone. We mention at this stage only that the set of real numbers is not

the same size as the set of natural numbers (though it is the same size as the set of points in the plane and the set of points in three-dimensional space).

Infinite Numbers and Cantor's Continuum Problem

Once you have grasped the concept of same size, you can go straight ahead and develop a system of 'numbers' that may be used to measure the 'size' of any set, be it finite or infinite. The 'numbers' themselves will be just abstractions, of course. The important point is that if two sets have the same size (i.e. if their elements can be paired off as described in the last section), then their 'size' (i.e. the 'number' of elements in each set) should be the same. So, for instance, when you measure the 'size' of the two sets

$$\{a, b, c\}, \qquad \{\text{Fred, Elsie, Fido}\},$$

you find that both sets have the same 'number' of elements, namely three. Likewise, when you measure the 'size' of the two infinite sets

$$\{1, 2, 3, 4, 5, \ldots\}, \qquad \{2, 4, 6, 8, 10, \ldots\},$$

you find again that these sets have the same 'number' of elements – in this case that 'number' being the smallest of the infinite numbers, denoted (following Cantor) by the symbol $\aleph_0$. (This is read as 'aleph-null'; aleph is the first letter of the Hebrew alphabet. The reason for the subscript zero will become clear in a moment.)

So what *is* the 'number three'? That which all sets of three elements have in common. Or, to put it another way, it is that which is common to all sets having the same size as the set $\{a, b, c\}$. Thus 3 is an abstraction which comes out of the notion of same size. There are various mathematical ways of making this statement precise, none of which will be gone into here. The main point is that finiteness is not at all relevant. So if you are happy (*were* happy?) with the concept of 'the number three', you should have no less ease in accepting the 'number' $\aleph_0$. It is that which is common to all sets having the same size as the set of all positive whole numbers.

As mentioned earlier, not all infinite sets have the same size – there is a whole (infinite) hierarchy of infinities. So, just as there is an infinite list of finite numbers, 1, 2, 3, ... , so too there is an infinite list of infinite numbers, $\aleph_0, \aleph_1, \aleph_2, \aleph_3, \ldots$, each one 'bigger' than the one before.

Addition and multiplication of Cantor's alephs turn out to be particularly simple (if at first sight a bit surprising). In each case the result is just the larger of the two infinite numbers. So, for example,

$$\aleph_0 + \aleph_1 = \aleph_1,$$
$$\aleph_1 \times \aleph_3 = \aleph_3.$$

(Hilbert's Hotel corresponds to the fact that $\aleph_0 + 1 = \aleph_0$. In order to cause an overflow from the hotel, $\aleph_1$ guests would have to arrive.)

Many of the infinite sets that occur in mathematics have size $\aleph_0$. For instance, the set of positive whole numbers, the set of all (i.e. positive and negative) whole numbers, the set of all rationals, and the set of all prime numbers all have size $\aleph_0$. But, as Cantor showed, the set of all real numbers very definitely has more than $\aleph_0$ members. Which at once raises the question, just what *is* the size of this set? Since it is not $\aleph_0$, it must be one of $\aleph_1, \aleph_2, \aleph_3, \ldots$, but which one? Despite many attempts, Cantor himself was unable to answer this seemingly simple question. So too were a number of other excellent mathematicians. Indeed, *Cantor's continuum problem*, as the question became known, resisted so many attempts at solution that when David Hilbert gave a keynote address at the International Congress of Mathematicians in Paris in 1900, he included the problem in a list of what he saw as the most significant challenges facing mathematicians as the new century dawned. (The name given to the problem arises because it asks for the size of the real *continuum* – this being the word used to describe the set of real numbers when considered as the points that make up the *real line*.)

Some progress was made when, in 1938, Kurt Gödel used new techniques of mathematical logic to show that from the Zermelo–Fraenkel axioms it is definitely not possible to prove that the set of real numbers does not have size $\aleph_1$. But this did not solve the problem, since it was still possible that the axioms were simply not sufficient to decide the issue one way or the other.

Despite this possibility, however, in the years following Gödel's result there was a widely held expectation that the continuum problem was in fact decidable within the Zermelo–Fraenkel framework. In which case, since Gödel had shown that the answer was definitely not provable to be anything other than $\aleph_1$, it had (it was assumed) to be the case that the continuum did have size $\aleph_1$, and that given time this would eventually be proved conclusively. Accordingly, it did not seem at all unreasonable to assume this anticipated result whenever a piece of mathematics required a knowledge of the size of the continuum, and a great many results were proved on the assumption that the *continuum hypothesis* (as the assumption was known) was true.

And then, in 1963, came the announcement that Paul Cohen of Stanford University had developed a new logical technique by which he had been able to *prove* that the continuum hypothesis could not be deduced from the Zermelo–Fraenkel axioms. When combined with the earlier result of Gödel, this showed that the continuum hypothesis was in fact *undecidable* in the Zermelo–Fraenkel system.

So what next? There are two ways of regarding the state of affairs brought about by Cohen's result. One conclusion is that it demonstrates the inadequacy of the axioms of Zermelo and Fraenkel. From this viewpoint the inadequacy is clearly a very severe one. It is one thing to know that the system is deficient as predicted by Gödel's incompleteness theorems, but for the system to be unable to resolve such a basic question as 'How many real numbers are there?' strikes a devastating blow at the theory. Some mathematicians reacted to this blow by suggesting that additional axioms should be formulated to overcome the new-found deficiency. But if you take that route, you are faced with the task of having to find appropriate additional axioms. Because of the essentially simple nature of set theory and its foundational position within mathematics, any axioms you introduce will have to be 'believable' – which means that even if they are not immediately obvious (and some of the Zermelo–Fraenkel axioms require some thought in order to see what they are saying), they will have to appear natural once you start to study them. (This consideration prevents you from taking the easy way out, of simply adopting the continuum hypothesis itself as an axiom of set theory – what justification is there for doing so?) The fact that mathematicians have been working with axiomatic set theory for over half a century now without encountering any such additional 'principle' leads most experts to conclude that there is in fact no 'missing axiom'.

So what you are left with is the alternative conclusion to be drawn from Cohen's discovery: that, however unpalatable it may seem, there simply is not one set theory but several. (Just as the nineteenth century brought the realization that there is not a single 'correct' geometry but three, alternative geometries, each with its own distinctive properties and results.) In some set theories the continuum hypothesis will be true, in others it will be false.

Did you notice the use of the phrase 'several set theories' above, not just 'two'? For it is not just the continuum hypothesis which is undecidable from the Zermelo–Fraenkel axioms. Following Cohen's initial discovery in 1963, it became apparent that his new method (the method of *forcing*) was applicable in a great many situations, not only in set theory itself. The two decades that followed saw the demonstration of the undecidability of a great many classical unsolved problems of mathematics. Gone for ever was the old expectation that, given enough time and ingenuity, any 'genuine' problem of mathematics could be resolved one way or the other. Besides the true statements and the false statements there is a third class of *undecidable* statements, statements which are neither true nor false. But at least Cohen's method of forcing provided a means of demonstrating that a problem belonged to this third class, so his result did make some positive contribution to mathematics.

Cantor's Proof

Just how did Cantor prove that there is a whole hierarchy of infinities? For readers who would like to see an example of a piece of highly abstract mathematical reasoning, here is a modernized version of Cantor's argument. It starts with the set-theoretic notion of a *subset*.

If X is any set, any collection of objects taken from X is called a *subset* of X. Thus, the collection

$$\{a, c, d\}$$

is a subset of the collection

$$\{a, b, c, d, e, f\},$$

and the set of prime numbers is a subset of the set of all integers.

Consider now the set consisting of *all* the subsets of the set X. Does such a set exist? Memories of the Russell paradox should be enough to warn you that you have to be careful about postulating the existence of sets. In this case there is no problem (as far as anyone knows). One of the axioms of Zermelo–Fraenkel set theory guarantees that there is such a set: it is called the *power set* of X, and is denoted by $P(X)$. For example, if $X = \{a, b\}$, then $P(X)$ consists of the sets

$$\emptyset, \{a\}, \{b\}, \{a, b\}.$$

What is that symbol $\emptyset$? It denotes the *empty set* (or *null set*), the set with no members. *If* it is indeed a set, then it will surely be a subset of any other set, since, in a trivial but none the less logically valid way, a set with no members has the property that 'all its elements' lie in any set X you choose. On the basis of which piece of reasoning alone you may be tempted to think that it is not a good idea to include amongst the other 'fictions' of mathematics the notion of an empty set. But then, by the same token, zero is not a sensible number. And there you have some idea as to why the empty set *is* taken as a bona fide set, along with all the other sets in mathematics. It is a 'zero element' much like the number 0. (Indeed, 0 is the number of elements in the set $\emptyset$.)

Another example: if $X = \{1, 2, 3\}$, then $P(X)$ is the set whose members are the sets:

$$\emptyset, \{1\}, \{2\}, \{3\}, \{1, 2\}, \{1, 3\}, \{2, 3\}, \{1, 2, 3\}.$$

The above two examples should already be enough to indicate that $P(X)$ seems to be a much larger set than X. When X has two members, $P(X)$ has four, and when X has three members, $P(X)$ has eight. In fact a fairly straightforward mathematical proof shows that if a finite set X has n members, then $P(X)$ has 2^n elements. You will appreciate from the discussion of the exponential function 2^n in Chapter 1 how the size of $P(X)$ increases very rapidly as more elements are added to X. It turns out that this difference in growth rates carries over into the infinite (though, as Hilbert's Hotel demonstrated, such carry-overs from the finite to the infinite should never be expected as a matter of course). This is how Cantor proved that there are infinitely many infinities. He demonstrated that for any set X, $P(X)$ is definitely not the same size as X (and hence is of a greater order of magnitude than X). The infinitude of the infinities follows easily from this. For suppose X is the set of natural numbers. Then $X_1 = P(X)$ is a set of greater infinite magnitude than X. Again, $X_2 = P(X_1)$ is larger than X_1. Similarly, $X_3 = P(X_2)$ is bigger than X_2, and so on.

In order to prove Cantor's result, suppose for the sake of argument that $P(X)$ were the same size as X. Our aim is to deduce a contradiction from this assumption. Then, provided the argument used is logically sound, the inescapable conclusion will be that the initial assumption must be false (since a true assumption cannot lead to a false or contradictory result). So here goes. Since X and $P(X)$ are assumed to be the same size, there will be a matching between these two sets. That is, for each member x of X there will be an associated member *partner-of-x* in $P(X)$, which is not matched to any other member of X; moreover, every member of $P(X)$ will be the partner of some member of X. Because this argument is intended to be valid for any set X, finite or infinite, there is no possibility of representing this matching using arrows as we did earlier – but see Box A.

Now consider an arbitrary member x of X. Then its partner, call it A, is a member of $P(X)$, i.e. A is a subset of X. So A consists of some of the members of X. Is x itself amongst these elements, i.e. is x a member of A? This is a perfectly reasonable question. For some members x of X the answer will presumably be 'yes', for others 'no'. Let U be the set of all those members x of X for which x is not a member of partner-of-x. (Does this remind you of anything? See below.) The set U consists of elements of X, and hence U is a subset of X, i.e. U is a member of $P(X)$. Thus U will be the partner of some member of X, call it w. (So $U = $ partner-of-w.)

Now ask the question: 'Is w a member of U?' If it is, then w must satisfy the requirement defining U, which is to say that w is not a member of its partner (in this case U). On the other hand, if w is not a member of U, then w fails to satisfy the requirement defining U, so w will be a member of its partner, U. This is an untenable situation – a contradiction. As mentioned earlier, the only possible conclusion is that the initial assumption that X and $P(X)$ were the same size must have been false. Thus X and $P(X)$ cannot be the same size, and Cantor's result is proved.

This is an excellent example of how a disaster can be turned to great advantage. The similarity between the above argument and Russell's paradox cannot have escaped you. But in this case all the steps can be justified from the Zermelo–Fraenkel axioms, and instead of the contradiction resulting in a paradox which destroys the entire theory, what you get is the desired falsity of the initial supposition.

And there you have it. The development of set theory has been one of the great success stories of modern mathematics, and there is nowadays scarcely any branch of mathematics that is not influenced to a greater or lesser extent by set-theoretic ideas and methods. Bertrand Russell once praised Cantor's achievement in giving birth to this subject as 'possibly the greatest of which the age can boast', and David Hilbert said, 'From the paradise created for us by Cantor, no one will drive us out.' Few would argue with these opinions.

Box A: *Proof of Cantor's theorem*

Given an infinite set X, the proof demonstrates that $P(X)$ does not have the same size as X (and hence is bigger). The idea is to show that the supposed existence of a matching between the elements of X and those of $P(X)$ leads to a contradiction. If the members of X are denoted by letters of the alphabet (a finite set, of course, but good enough for this illustration), then part of such a supposed matching might look like this:

Element of X		*Subset of X*
y	$\longleftrightarrow$	$\{a, b, c, d\} = A(y)$
z	$\longleftrightarrow$	$\{b, d, p, q, z\} = A(z)$

In this case y is not a member of the set $A(y)$ with which it is matched, whereas z is a member of its matched set $A(z)$. Let U denote the set of all those members of X which are not members of their matched set:

$$U = \{x \mid x \notin A(x)\}.$$

Since U is a subset of X, U must be matched with some element of x, say w:

$$w \longleftrightarrow U = A(w).$$

Investigation of whether or not w is a member of the set $A(w)$ now leads to a contradiction. If w is a member of $A(w)$, then this means that w cannot be in U, but since U is just $A(w)$ this is contradictory. On the other hand, if w is not a member of $A(w)$, then w will be a member of U, and again, since $U = A(w)$, there is a contradiction.

Suggested Further Reading

For an introductory account of mathematical logic there is *What is Mathematical Logic?*, by J. N. Crossley (Oxford University Press, 1972) and *Language, Logic and Mathematics*, by C. W. Kilmister (English Universities Press, 1967), as well as a variety of other books. Wilfred Hodges' book *Logic* (Pelican, 1977) deals mostly with the non-mathematical side of the subject, but is worth looking at.

A gentle but thorough introduction to set theory which covers all the topics mentioned in this chapter is *Fundamentals of Contemporary Set Theory*, by Keith Devlin (Springer-Verlag, 1980).

3 Number Systems and the Class Number Problem

The Solution to a 180-year-old Problem

In 1983 Don Zagier of the University of Maryland and the Max Planck Institute in Bonn, and Benedict Gross of Brown University, Providence, Rhode Island, announced that they had solved the class number problem, a famous (among mathematicians) problem posed by Karl Friedrich Gauss in 1801. Though their proof was by no means the longest in mathematics (Chapter 5 deals with that), at 300 pages it was longer than most. But it is not the length of the proof that mathematicians find so fascinating; it is its nature. It is very indirect, and links two seemingly unrelated areas of mathematics in a quite remarkable way.

Though both the problem and its solution are highly abstract and involve some very difficult mathematics, at heart it is all to do with number systems of one kind or another, and it is certainly possible to describe the general issues involved. This is what this chapter sets out to do. Along the way we shall trace out a great deal of the historical development of present-day mathematics. But we start out by taking a look at:

The Remarkable Properties of the Number 163

In the eighteenth century the great Swiss mathematician Leonhard Euler discovered (no one knows how) that the formula

$$f(n) = n^2 + n + 41$$

has a rather remarkable property. If you put n equal to any of the numbers from 0 to 39, the resulting value of $f(n)$ is a prime number. For instance, $f(0) = 41$ is prime, as are $f(1) = 43$ and $f(2) = 47$. No other quadratic formula has been discovered which produces anything like as many prime numbers (starting from $n = 0$ and working up through successive values of n). And though the unbroken sequence of prime numbers stops with $n = 40$, when $f(40) = 41^2$, the formula still produces a lot of prime numbers. Of the first 10 million values the proportion of primes is about one in three – far greater than for any other quadratic formula. (See Chapter 6 for a further discussion of prime-generating formulas.)

Since Euler's formula seems so unusual in its production of primes, there is presumably something rather special about it. But what? Well, properties of formulas for integers often turn out to be closely related to properties of the same formulas regarded as formulas for real numbers (or even for complex numbers). Indeed, there is an entire branch of mathematics, known as *analytic number theory*, which exploits this phenomenon (see Chapter 9). What happens when Euler's formula is regarded as a formula for real numbers?

First, let us rewrite the formula with x (the usual symbol for a real number) in place of n (the usual symbol for an integer), thus:

$$f(x) = x^2 + x + 41.$$

Anyone who can remember the algebra they learnt at school ought to find this suggestive of *quadratic equations*: equations of the form

$$ax^2 + bx + c = 0,$$

which have to be solved for x when the values of a, b, c are known. You may even remember that there is a formula which gives the two solutions:

$$x = \frac{-b \pm \sqrt{b^2 - 4ac}}{2a}$$

(The two possibilities for the 'plus-or-minus' sign give two solutions to the equation.)

Because it is not possible to take the square root of a negative number (when real numbers are being considered), the sign of the expression $b^2 - 4ac$ is very important. If this expression is positive the quadratic equation will have two solutions; if it is negative there will be no (real) solutions. (And if it is zero then there will be only one solution – this is a special case.) The expression is called the *discriminant* of the quadratic equation.

What is the discriminant of the Euler quadratic

$$x^2 + x + 41?$$

Here $a = 1$, $b = 1$, and $c = 41$, so

$$b^2 - 4ac = 1 - 164 = -163.$$

As the discriminant is negative, we know at once that the quadratic equation

$$x^2 + x + 41 = 0$$

has no (real) solutions. (There are two complex solutions:

$$x = -\tfrac{1}{2} \pm \tfrac{1}{2}\sqrt{163}i.$$

See later in this chapter for a discussion of complex numbers.)

And there, believe it or not, lies the reason for the special behaviour of the Euler formula as a prime generator. Not that the discriminant is negative – lots of formulas have that property, but that its value is -163. 'What's so special about 163?' you might ask. Read on and you will discover that it is a very special number indeed, closely related to some fundamental mathematical constants.

What are the most common special 'constants' of mathematics, the numbers which keep cropping up in the most unexpected places? The most obvious one is π, the ratio of the circumference of a circle to its diameter.

(This definition already indicates that π is special. Why should you get the same answer for every circle, whatever its size?)

Since the late nineteenth century it has been known that π is *irrational*, which is to say that its decimal representation continues indefinitely, without settling into any repetitive pattern. To twenty decimal places,

$$\pi \ = \ 3 \cdot 141\,592\,653\,589\,793\,238\,46.$$

Computers have been used to calculate π to more than 30 million places.

Besides its definition in terms of circles, π occurs in many other situations. For instance, the sum of the terms in the infinite sequence

$$1 - \frac{1}{3} + \frac{1}{5} - \frac{1}{7} + \frac{1}{9} - \frac{1}{11} + \ldots$$

has the value $\pi/4$. (The development of methods for dealing with infinite sums such as this was one of the crowning achievements of nineteenth-century mathematics.) Again, the sum of

$$1 + \frac{1}{4} + \frac{1}{9} + \frac{1}{16} + \frac{1}{25} + \ldots$$

(where the nth term in the sequence is the reciprocal of n^2) is $\pi^2/6$.

Another surprising appearance of π is this: if you throw a matchstick onto a board on which are ruled parallel lines one matchstick-length apart, the probability that the matchstick will end up touching one of the lines is precisely $2/\pi$.

After π, the next most common mathematical constant is e, the base of the natural logarithms. Like π, the number e is irrational; its decimal representation is infinite. To twenty decimal places,

$$e \ = \ 2 \cdot 718\,281\,828\,459\,045\,235\,36.$$

And like π, there are various ways of defining e. For instance, e is that number for which the graph of the function

$$y \ = \ e^x$$

has the property that the gradient at any point equals the value (of y) at that point. For example, if a population p is growing according to the law

$$p = e^t,$$

(where t is the time) then the rate of growth at any instant is exactly equal to the population size at that instant.

Another (related) definition of e is that it is that number such that the area bounded by the lines $y = 1/x$, $y = 0$, $x = 1$, and $x = e$ is exactly equal to 1 (see Figure 3). Expressed in terms of an integral, this definition of e is such that

$$\int_1^e \frac{1}{x}\, dx = 1.$$

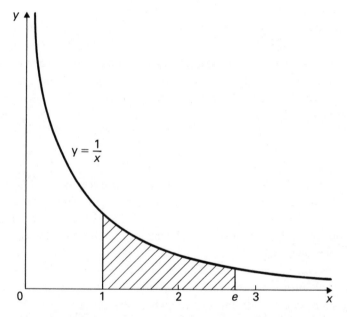

Figure 3. Definition of the mathematical constant e as the number such that the shaded area is exactly 1.

Still another definition involves an infinite sum:

$$e = 1 + \frac{1}{1!} + \frac{1}{2!} + \frac{1}{3!} + \frac{1}{4!} + \cdots ,$$

where N! (to be read as 'N factorial') denotes the product $1 \times 2 \times 3 \times \ldots \times N$. In fact this is just a special case of the formula

$$e^x = 1 + \frac{x}{1!} + \frac{x^2}{2!} + \frac{x^3}{3!} + \frac{x^4}{4!} + \ldots .$$

Anticipating for a moment the topic of complex numbers to be discussed later in this chapter, the above formula is valid even if the number x is *complex*; that is, if it has the form $a + bi$ where $i = \sqrt{-1}$. This leads to some amazing results. For example, Euler discovered that

$$e^{\pi i} = -1.$$

In other words, when the irrational number e is raised to the power of the irrational number π times the imaginary number $\sqrt{-1}$, the result is the whole number -1. Another equally startling result which relates e, π, and $\sqrt{-1}$ is

$$i^i = e^{-\pi/2} = 0 \cdot 207\,879\,576\,3 \ldots .$$

And now to the point of this discussion of mathematical constants. The three numbers π, e, and $\sqrt{163}$ are all irrational. And yet, to twelve decimal places,

$$e^{\pi\sqrt{163}} = 262\,537\,412\,640\,768\,744 \cdot 000\,000\,000\,000.$$

In point of fact, this number is not an integer. A more accurate value is

$$262\,537\,412\,640\,768\,743 \cdot 999\,999\,999\,999\,250,$$

which is correct to fifteen places of decimals. Nevertheless, the result is very nearly an integer, whereas for most other natural numbers k the value of $e^{\pi\sqrt{k}}$ will be nothing like as close. The surprising thing is that the value of k for which this occurs is that number 163 again. You might suspect that this is not just an accident, but that there is something going on behind the scenes. And you would be right. Just what it is that is so special about this number 163 will be revealed as the remainder of this chapter unfolds. The story begins in Ancient Greece.

Early Number Systems

The Ancient Greeks appear to have been the first to develop a mathematical theory of arithmetic. Both the Ionian School (founded by Thales around 600 BC) and the Pythagorean School (founded by Pythagoras some fifty years later) developed extensive theories of both geometry and (particularly the Pythagoreans) arithmetic. It was the Greeks who first recognized the *natural* (or *counting*) *numbers* 1, 2, 3, ... as forming an infinite collection on which the basic arithmetic operations of addition and multiplication could be performed. Though they did not acknowledge *negative* numbers as such, they knew how to manipulate minus signs in expressions such as

$$(7 - 2) \times (6 - 3) = (7 \times 6) - (7 \times 3) - (2 \times 6) + (2 \times 3).$$

As such, their approach was probably not unlike that summed up by the old schoolroom jingle:

> '*Minus times minus equals plus,*
> *The reason for this we need not discuss.*'

However, there was a good reason why the Greeks refused to countenance as a number an entity such as -5. To them, numbers were closely bound up with *measurement* – of distance, area, and volume. Their algebraic rules were generally thought of in geometric terms such as adding together various areas (see Figure 4).

But if they had no use for negative numbers, the Greeks certainly needed fractions, or *rational numbers* as the mathematician calls them. A (positive) *rational* number is one of the form a/b, where a and b are both natural numbers. Since b may be 1, the rational numbers include the natural numbers. (In the terminology of Chapter 2, the natural numbers form a *subset* of the rational numbers.) Until some point in the sixth century BC the Greeks believed that the system of (positive) rational numbers was adequate for their geometrical purposes. But then they discovered to their horror that this was not the case at all. In particular, it was found that the square root of 2 is not a rational number, which meant that the rational numbers are not adequate for measuring (say) the hypotenuse of a right-angled triangle

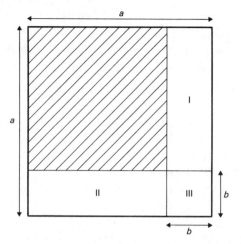

Figure 4. Greek algebra. The Greeks regarded familiar algebraic identities such as

$$(a - b)^2 = a^2 - 2ab + b^2$$

in purely geometric terms. To obtain the shaded area, i.e. $(a - b)^2$, you start with the entire square (a^2), subtract the rectangle consisting of regions I and III (ab) and that consisting of regions II and III (also ab) and then add on the small square III (b^2) to compensate for the fact that this area was included in *both* subtracted rectangles. This gives the identity above.

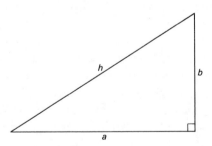

Figure 5. Pythagoras' theorem. For any right-angled triangle having sides a and b adjoining the right-angle and h the hypotenuse, the identity

$$h^2 = a^2 + b^2$$

holds. When $a = b = 1$, this identity gives $h = \sqrt{2}$, an irrational quantity, not expressible as a quotient of two whole numbers.

whose base and height both measure 1 unit (see Figure 5). (In order to be able to measure all geometric lengths the *real* numbers are required, of which more in a moment.) This discovery effectively marked the end of any progress in arithmetic made by the Greeks, who henceforth restricted their mathematics to geometrical construction.

Negative Numbers

The first systematic algebra to use zero and negative numbers was developed by Hindu mathematicians in the seventh century AD. They used positive and negative numbers to handle financial transactions involving credit and debit. Besides being the first to use zero in a modern fashion, they wrote equations with negative numbers symbolized by a dot above the number (an early forerunner of our minus sign) and explicitly formulated a *law of signs* (plus times plus is plus, plus times minus is minus, minus times minus is plus). They also recognized that every positive number has two square roots – one positive, the other negative.

However, these early developments in India did not affect the European mathematicians of the Renaissance period, from the fourteenth to the sixteenth century. Following the Greek tradition, they were happy to manipulate minus signs but did not recognize negative numbers as such. Negative roots to equations were referred to as 'fictitious roots'.

By the seventeenth century, some mathematicians were starting to use 'negative numbers', but the tendency met with opposition, sometimes from prominent mathematicians. René Descartes spoke of negative roots as 'false roots', and Blaise Pascal too thought that there could be no such thing as a number less than zero. Gottfried Leibniz, whilst agreeing that they could lead to absurdities, nevertheless defended negative numbers as a useful device in performing calculations. Leonhard Euler accepted negative numbers but believed they were greater than infinity (∞), reasoning that since $a/0 = \infty$, then if we divide a by a number smaller than zero the result must be greater than infinity.

It was during the eighteenth century that the algebraic use of negative numbers (denoted by the minus sign) finally became widespread, though even then many mathematicians still felt uneasy about them and would often go to great lengths to avoid their use if it were at all possible. Indeed, it is only when you adopt the axiomatic approach to numbers (see Chapter 2) that negative numbers really 'make sense'. This remark applies equally well to the complex numbers, but before we can look at them we ought to say something about the 'real' numbers.

Real Numbers

Though the present discussion of number systems is divided into sections according to the different types of number, historically there was no such distinction, and the theories of negative numbers, real numbers, and complex numbers all evolved over roughly the same period. Of the three, a proper treatment of the real numbers was by far the greatest accomplishment (being by far the most difficult), and although (as will be indicated) in the theory of complex numbers the prior existence of the real number is assumed, it was the theory of the real numbers that was the last to be worked out.

For all *practical* purposes, the rational numbers are more than sufficient. Indeed, in the real world (as opposed to the mathematical one) these are the only numbers that are used, with answers to problems being given to at most a few decimal places. But of course, the rational numbers also possess some pleasant *mathematical* properties. If you add, subtract, multiply, or divide (except by zero) two rational numbers, the result is again a rational number. Moreover, the arithmetic of rational numbers satisfies all the axioms for an integral domain that were listed in Chapter 2 (pp. 31–2). The mathematician would sum all this up by saying that the rational numbers constitute a *field*. What is a *field*? It is an integral domain in which division is possible; i.e. a structure satisfying the seven axioms on pp. 31–2 together with the additional axiom:

(8) For any number x other than 0, there is a number y such that $xy = 1$. (Existence of multiplicative inverses.)

It is an easy matter (try it) to verify that the y guaranteed to exist by Axiom 8 is unique for any given x. Normally we write x^{-1}, or sometimes $1/x$, to denote this unique number y. Axiom 8 enables you to divide because, of course, a/b is the same as ab^{-1}.

Briefly then, a *field* is a structure which allows you to perform all the usual arithmetical operations with all the usual properties. Where the rational number field falls down is in its inability, as discovered by the Pythagoreans, to allow the solution of equations such as

$$x^2 = 2.$$

Using rational numbers, you can find a solution to any prescribed degree of accuracy:

$$1^2 = 1, \quad (1{\cdot}4)^2 = 1{\cdot}96, \quad (1{\cdot}41)^2 = 1{\cdot}9981,$$
$$(1{\cdot}414)^2 = 1{\cdot}999\,396,$$

and so on. But there is no rational number whose square is *exactly* equal to 2. The *real* numbers, on the other hand, constitute a field which includes the rational numbers and is rich enough to solve equations like the one above. The key idea in formulating this in a precise way is provided by the process of *successive approximation* inherent in the above example. The numbers 1, 1·4, 1·41, 1·414, ... provide better and better approximations to a 'number' whose square is 2. If it were possible to employ infinitely many places of decimals, then we would be able to write down a number whose square was exactly equal to 2, namely

$$1{\cdot}414\,213 \ldots \ (ad\ infinitum).$$

But since we obviously cannot write down an infinite sequence of decimal places, how do we proceed in practice? Well, by allowing the mathematics to handle the necessary infinite concepts for us, which means that the real numbers have to be developed in an axiomatic way. It turns out to be an extremely difficult process, far outside the scope of a book such as this. Indeed, the formulation of a proper axiom system for the real numbers was one of the greatest accomplishments of mathematics when it was finally achieved during the 1870s.

The real numbers include all the rational numbers (just as the integers form a subset of the rational numbers), but a great many other numbers besides. A real number which is not rational is called an *irrational number*. Examples of irrational numbers are π, e, and $\sqrt{k}$ for any natural number k which is not a perfect square.

Complex Numbers

During the sixteenth century, European mathematicians – and in particular the Italian Rafaello Bombelli – began to realize that in the solution of algebraic problems it was often useful to assume that negative numbers have square roots. It is perhaps understandable when we consider the climate of the time that such 'numbers' were referred to as *imaginary numbers*, though to the present-day mathematician *all* numbers are 'imaginary' concepts, the square roots of negative quantities no more or no less so than any others. Nevertheless it is still customary to refer to a square root of a negative real number as *imaginary*, thereby giving the word 'imaginary' a special, technical meaning in this context.

In fact, in order to have available the square roots of negative real numbers it is necessary only to postulate the existence of a solution to the one equation

$$x^2 + 1 = 0.$$

If i denotes a solution to this equation (so $i^2 = -1$) then, for any positive real number a, the square root of the negative number $-a$ will be $(\sqrt{a})i$. (Actually there will be two square roots, $(\sqrt{a})i$ and $-(\sqrt{a})i$. Likewise there will be two solutions to the equation $x^2 + 1 = 0$, namely i and $-i$.) It is the numbers of the form ai, where a is real, that are called *imaginary numbers*. (The letter i was first used in this context by Leonhard Euler.)

A *complex number* is one of the form $a + bi$, where a and b are real numbers. The plus sign here is not intended to denote ordinary addition (how could it?), rather it serves to separate the *real part a* of the complex number from its *imaginary part bi*. Notice that if $b = 0$, then $a + bi = a$, so the real numbers form a subset of the complex numbers. Similarly, if $a = 0$, then $a + bi = bi$, so the imaginary numbers also form a subset of the complex numbers.

At this stage you might be thinking that there is no justification for calling something of the form $a + bi$ a number, even if you are prepared to countenance $i = \sqrt{-1}$ in the first place. But remember, it is not what numbers *are* that matters, but how they *behave*. Provided the complex numbers have a workable and useful (either in mathematics itself or possibly in a wider context) arithmetic, possibly forming a field, then they have as much right to be called 'numbers' as do any others. So what is the arithmetic of complex numbers? The rules are given below. (And for most people this is the first number system which they actually *meet* from an axiomatic viewpoint. By the time any of the integers, the rational numbers, or the real numbers are developed axiomatically, most people are already quite familiar with them.)

The rule for adding two complex numbers is quite simple: you add their real parts together, and you add their imaginary parts together. Thus,

$$(a + bi) + (c + di) = (a + c) + (b + d)i.$$

So, for example,

$$(2 + 3i) + (7 + 1i) = 9 + 4i,$$
$$(-3 + 4i) + (4 - 2i) = 1 + 2i.$$

Multiplication of complex numbers is a little more complicated. The idea is to use the ordinary rules of algebra for multiplying two bracketed sums and then putting $i^2 = -1$. Thus,

$$\begin{aligned}(a + bi)(c + di) &= ac + adi + bci + bdi^2 \\ &= ac + adi + bci - bd \\ &= (ac - bd) + (ad + bc)i.\end{aligned}$$

So, for example,

$$\begin{aligned}(2 + 3i)(5 + 7i) &= 10 + 14i + 15i + 21i^2 \\ &= 10 + 14i + 15i - 21 \\ &= -11 + 29i\end{aligned}$$

Perhaps surprising is that complex numbers can be divided. The rule is:

$$\frac{a + bi}{c + di} = \frac{ac + bd}{c^2 + d^2} + \frac{bc - ad}{c^2 + d^2}i.$$

So, for example,

$$\frac{3 + 5i}{1 + 2i} = \frac{3 \times 1 + 5 \times 2}{1 + 4} + \frac{5 \times 1 - 3 \times 2}{1 + 4}i$$

$$= \frac{3 + 10}{5} + \frac{5 - 6}{5}i$$

$$= \frac{13}{5} - \frac{1}{5}i.$$

In fact the complex numbers form a field. (As an exercise you might like to check that the definitions of addition and multiplication given above do lead to an arithmetic which satisfies all the field axioms.) So however strange you may feel the very notion of a complex number to be, it does turn out to provide a 'normal' type of arithmetic. In fact it gives you a tremendous bonus not available with any of the other number systems. In the complex-number field *every* polynomial equation can be solved! That is, if $a_0, a_1, \ldots, a_{n-1}, a_n$ are complex numbers, then there will be a complex number x which is the solution of the equation

$$a_n x^n + a_{n-1} x^{n-1} + \ldots + a_1 x + a_0 = 0.$$

This is not true of the real numbers, of course, as the equation $x^2 + 1 = 0$ testifies.

The result just mentioned is known as the *fundamental theorem of algebra*. It was first stated by Girard in 1629, and then proved imperfectly by d'Alembert in 1746 and Euler in 1749. The first totally correct proof was provided by Gauss in his 1799 doctoral thesis. So impressed was Gauss with the result that he subsequently gave three further (quite distinct) proofs of it.

The fundamental theorem of algebra is just one of several reasons why the complex-number system is such a 'nice' one. Another important reason is that the field of complex numbers supports the development of a powerful differential calculus, leading to the rich theory of functions of a complex variable. (This theory is touched upon in Chapter 9.)

Not only is the theory of complex numbers appealing mathematically, it turns out to be extremely useful as well. The first significant scientific use of complex numbers was made by Charles Steinmetz, who found them essential for performing efficient calculations concerning alternating currents. Indeed, nowadays no electrical engineer could get along without

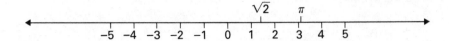

Figure 6. The real line. The axioms for the real numbers guarantee that the line is 'continuous', having no 'holes', not even infinitely small holes where a single point is left out (in the sense that the 'rational line' has a 'hole' where $\sqrt{2}$ ought to be).

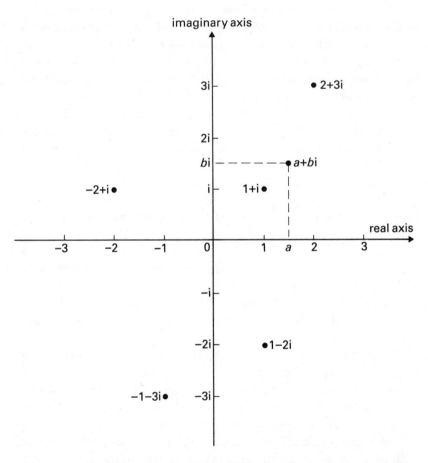

Figure 7. The complex plane. The complex number $a + bi$ is identified with the point having coordinates (a, b). The real numbers lie along the horizontal axis, and the pure imaginary numbers on the vertical axis.

complex numbers, and neither could anyone working in aerodynamics or fluid dynamics. In Einstein's theory of relativity use is made of complex numbers: the three spatial dimensions are regarded as real and the time dimension as imaginary; and in quantum mechanics the physicist has to deal with complex numbers.

And yet, *despite* the fact that they constitute a field, *despite* the fact that they are very useful, and *despite* the fact that all number systems are purely abstract, 'imaginary' constructs, many people still feel uneasy with the complex numbers. It is largely (entirely?) a matter of familiarity. For instance, the real numbers may turn out to be extremely complicated mathematical objects when put under the analyst's 'microscope', but there is always the comfortingly simple picture of the *real line* to fall back on – a 'two-way infinite' straight line with 0 at the middle (see Figure 6).

The good news is that there is an equally comforting picture of the complex numbers. Just as the real numbers can be thought of as the points on the real line, so the complex numbers can be identified with the points in a two-dimensional plane (see Figure 7). The first person to propose this visualization of complex numbers was Caspar Wessel, a self-taught Norwegian surveyor who lectured on his idea in 1797. The same idea was rediscovered by Jean-Robert Argand, a Swiss bookkeeper, who published a book on the subject in 1806, and also by Gauss. Argand's book proved to have the greatest initial impact, and the *complex plane*, as the two-dimensional plane is properly called when intended as a representation of complex numbers, is sometimes still referred to as the *Argand diagram*.

Quaternions

Inspired by the representation of complex numbers as points on a plane, the Irish mathematician William Rowan Hamilton (1805–65) developed an algebraic (i.e. essentially axiomatic) interpretation of complex numbers in terms of pairs of real numbers. He went on to investigate the possibility of a three-dimensional analogue of the complex number plane. This proved impossible but, as Hamilton discovered, when you go to four dimensions it is possible to develop a system of what might be called 'hyper-complex numbers'.

The *quaternions*, as Hamilton called his new numbers, did not come easily, and it was only after several years' thought that he was able to make the crucial breakthrough. As so often in mathematical research, the key insight did not come to him whilst he was seated at his desk. At dusk one day in 1843, he was strolling with his wife along the Royal Canal in Dublin, when he realized that provided he dropped the commutative law for multiplication, everything else would work (i.e. he would get an otherwise acceptable number system). He was so elated at this insight that he stopped at Brougham Bridge to scratch the basic formulas on a stone. (The original graffito has long since weathered away, but the bridge now bears a plaque commemorating the great event.)

Briefly, a quaternion is a number of the form

$$a + bi + cj + dk$$

where a, b, c, d are real numbers, and i, j, k are 'imaginary' numbers which satisfy the equation $i^2 = j^2 = k^2 = -1$. The key equations that Hamilton inscribed on the bridge are:

$$ij = k, \quad jk = i, \quad ki = j,$$
$$ji = -k, \quad kj = -i, \quad ik = -j.$$

By using these rules, any two quaternions may be multiplied together (to give a third quaternion) by means of the ordinary rules of algebra. (Addition is simply term by term, as with complex numbers.) The resulting system of numbers satisfies all the axioms for an integral domain (pp. 31–2) except for the commutative law of multiplication.

Quaternions have found considerable use in modern physics, as has an even more bizarre number system – the *octonions*, an eight-dimensional number system in which not only the commutative law of multiplication, but the associative law too is lost. But now it is time to go back to the natural numbers, and in particular to Gauss's work in number theory.

The Gaussian Integers

In 1796 Gauss proved a profound theorem of number theory called the *quadratic reciprocity law*, which concerns the solution of equations such as

$$x^2 \bmod 7 \; = \; 3,$$

which are of the form

$$x^2 \bmod p \; = \; q,$$

where p and q are prime numbers. In trying (with eventual success) to generalize his theorem to cover equations of higher order ($x^3 \bmod p = q$, and so on) he found that his calculations were made easier by working with numbers of the form $a + bi$, where a and b are integers (and $i = \sqrt{-1}$, as usual), rather than with the integers alone. Nowadays such 'complex integers' are known as *Gaussian integers*. They are particularly useful when factorization is involved. Just as the ordinary integers allow the factorization

$$a^2 - b^2 \; = \; (a + b)(a - b),$$

so the Gaussian integers give

$$a^2 + b^2 \; = \; (a + bi)(a - bi).$$

Intuitively, the Gaussian integers would appear to occupy the same position within the complex-number field as do the ordinary integers within the real-number field. But just how like the integers are the Gaussian integers?

As was made clear in Chapter 1, the most significant fact about the integers is encompassed by the *fundamental theorem of arithmetic*: that every integer is expressible as a product of a unique collection of primes (numbers which cannot be split up any further), and possibly -1. Gauss demonstrated that amongst the Gaussian integers are certain numbers which are 'prime' (i.e. cannot be split up), and that in terms of these 'primes', an analog of the fundamental theorem of arithmetic holds for the Gaussian integers: the *unique factorization theorem*. (The 'primes' here are not numbers of the form $a + bi$, where both a and b are primes: the Gaussian primes are defined as being those Gaussian integers which cannot be reduced to a product of other Gaussian integers. For this reason mathematicians often refer to them as *irreducibles*.)

The Class Number Problem

The Gaussian integers turned out to be useful in other contexts besides reciprocity laws – notably their connection with Fermat's last theorem, of which more in Chapter 8. So useful did they prove to be that it made sense to examine other, similar number systems. This is what Gauss did. Of the various approaches that can be made, particularly fruitful are the systems of the form $a + b\sqrt{-d}$, where d is some positive integer other than 1.

At this stage there is a minor surprise in store. In order to obtain a reasonable number system (i.e. one that bears some resemblance to the ordinary integers), in the case where $d \bmod 4 = 3$ you have to allow the numbers a and b to be half-integers as well as integers. So, for example,

$$\frac{1}{2} + 2\sqrt{-3}, \qquad \frac{3}{2} + \frac{5}{2}\sqrt{-3}$$

will be numbers in the system which corresponds to $d = 3$. (If $d \bmod 4 \neq 3$ then, as with the Gaussian integers, a and b are just integers.)

Once the above minor modification has been dealt with, you can ask for which values of d you get a reasonable 'number theory'. In particular, for which values of d do you get a unique factorization theorem? For $d = 1$ (the Gaussian integers) you do. So too for $d = 2$ and $d = 3$. But for $d = 5$ you do not. In this system the number 6 (for example) has the two factorizations (into irreducibles)

$$6 = 2 \times 3, \quad 6 = (1 + \sqrt{-5}) \times (1 - \sqrt{-5}).$$

In Gauss's time, nine values of d were known for which the system of numbers $a + b\sqrt{-d}$ (for a and b varying as indicated above) has a unique factorization theorem. They are

$$d = 1, 2, 3, 7, 11, 19, 43, 67, 163.$$

Are there any more values? Despite considerable efforts by Gauss and others in the decades that followed, no one was able to find any. The next result was the discovery by Heilbronn and Linfoot in 1934 that there could be *at most* one extra value, and if it existed it would have to be astronomically large. But *was* there a tenth value?

In 1952 one man knew that there was not. In that year Kurt Heegner, a retired Swiss scientist who did mathematics as a hobby, published what he claimed to be a proof that there was no tenth d, but no one believed him. (His paper was *very* hard to follow. Even so ... .) The rest of the world had to wait another fifteen years before they knew the truth. In 1967, Harold Stark of the Massachusetts Institute of Technology and Alan Baker of the University of Cambridge independently (and using different methods) also proved that there was no tenth d, and this time the mathematical community was convinced. Motivated by their discovery, Stark and Baker set about looking at Heegner's earlier work, and to their amazement found that it was essentially correct. The neglected Swiss had been right after all!

And there you have the reason why that number 163 is so special, giving rise to those curious results mentioned at the beginning of this chapter. It is the largest value of d for which the number system $a + b\sqrt{-d}$ allows unique factorization. (Unfortunately it is not possible to give any indication here of just how the unique factorization property for 163 relates to the earlier discussion of that number. That is strictly for the professional mathematician!)

Having disposed of those number systems $a + b\sqrt{-d}$ which do allow unique factorization, what can be said about the ones that do not? Once again, Gauss led the way. With each number system derived from some value of d he associated a certain natural number $h(d)$ called the *class number* of that system. This class number gives a measure of the margin by which unique factorization fails. If the class number is 1 (which it is for each value of d in Gauss's list), then unique factorization holds. If $h(d) = 2$ (as it is for $d = 5, 6, 10, 13$, for example) then unique factorization just fails. For class number 3 (for $d = 23, 31, 59$, for example) it fails a little more, for class number 4 (for $d = 14, 17, 21$, for example) it fails still more, and so on. The bigger the class number, the more ways there are of factoring numbers in the system into 'primes' (of the system).

In Article 303 of his *Disquisitiones Arithmeticae* (Gauss's monumental work, mentioned in Chapter 1, Gauss described some extensive computations of class numbers, and observed that for each class number k there seemed to be a largest value of d for which $h(d) = k$. The largest d for which $h(d) = 1$ was (as far as he knew) $d = 163$, the largest d with $h(d) = 2$ seemed to be $d = 427$, and the largest d with $h(d) = 3$ was apparently 907. But Gauss was able neither to confirm that any of these values really was the largest, nor to prove that there always was a largest d, though he conjectured that this was nevertheless the case.

The *class number problem* (which assumes the truth of Gauss's conjecture in order to make sense) is to determine for each class number k the largest d for which $h(d) = k$. (Thus Heegner's 1952 result solved the class number problem for the case $h = 1$.)

Virtually no progress was made on the class number problem from Gauss's time until the present century. In 1916 Hecke proved that if a certain rather complicated statement known as the *generalized Riemann hypothesis* were true, then Gauss's conjecture that each class number corresponded to only finitely many values of d would follow. But since no one had (nor indeed *has*) any idea of whether the generalized Riemann hypothesis is true or not, Hecke's result did not say very much. Or at least not on its own it didn't. But then in 1934, building on recent work by Deuring and by Mordell, Heilbronn proved Gauss's conjecture on the assumption that the generalized Riemann hypothesis was *false*. Since the hypothesis in question will certainly have to be true or false – even if we do not (or, bearing in mind the results described in Chapter 2, *cannot*) decide which – the results of Hecke and Heilbronn taken together finally *proved* Gauss's conjecture.

With Gauss's conjecture finally established, the way was clear to try to solve the class number problem itself. But progress proved to be painfully slow. First there was Heegner's 1952 result for the case $h = 1$. Then, in 1967, as well as re-solving the case $h = 1$ Baker and Stark disposed of the case $h = 2$. But none of the methods that had been developed was able to handle any other cases.

The big breakthrough came in 1975, when a 'partial' solution was obtained by Dorian Goldfeld of the University of Texas at Austin. By means of a long and difficult argument in analytic (complex) number theory, Goldfeld showed that, provided a certain (rather complex) mathematical object were available, the entire solution to the class number problem would follow. The object he required was a geometric curve of a certain shape,* having some unusual special properties. Finding curves of the requisite shape was not the problem; getting one with the special properties was. Goldfeld tried hard to find one and failed, as did everyone else who looked at the problem.

Until, in 1983, Zagier and Gross were successful in their search. Their key idea was to look for certain special points on the curve, and in honour of the long-neglected Heegner they call these special points *Heegner points*. What the proof then amounted to was one enormous equation. Simply calculating the two sides of the equation took up 100 pages. Then they had to pair off terms on each side to prove that the equation was correct! But despite its length, this part of the proof is what mathematicians refer to as 'straightforward'. What is really remarkable is the fact that a single curve somehow controls the behaviour of an infinite family of number systems.

After 183 years, Gauss's class number problem had finally been put to rest.

Suggested Further Reading

For a historical account of the development of number systems see *Numbers: Their History and Meaning*, by Graham Flegg (André Deutsch, 1983).

A brief description of number systems is provided by the book *Numbers: Rational and Irrational*, by Ivan Niven (Random House, 1961). A more complete coverage can be found in *Foundations of Real Numbers*, by Claude Burrill (McGraw-Hill, 1967) and also in *The Structure of the Real Number System*, by Leon Cohen and Gertrude Ehrlich (Van Nostrand, 1963).

For a nice account of the kinds of number system that figure in the class number problem, see Chapter 8 of *An Introduction to Number Theory*, by Harold Stark (Markham, Chicago, 1970). At a higher level there is *Algebraic*

---

*In particular, the curve should have an equation of the form

$$y^2 = ax^3 + bx^2 + cx + d.$$

Such curves are called elliptic curves. They have a number of applications in number theory besides the one described here.

Number Theory, by I. N. Stewart and D. O. Tall (Chapman and Hall, 1979).

A *very* high-level account of the final solution to the class number problem is provided by Don Zagier's article '*L*-series of elliptic curves, the Birch–Swinnerton–Dyer conjecture, and the class number problem of Gauss', which appeared in the mathematical journal *Notices of the American Mathematical Society*, Volume 31, Number 7 (November 1984), Issue 237, pp. 739–43. This article also provides further (high-level) sources which can be followed up by those who wish to do so. But it should be stressed that this work is very advanced, and the majority of professional mathematicians would find it hard going trying to understand the proof.

Finally, if you want to take a look at Gauss's *Disquisitiones Arithmeticae* yourself, an English version was brought out by Yale University Press in 1966 (the original was published in Leipzig in 1801).

4   Beauty From Chaos

Beauty in Mathematics

It was Bertrand Russell who wrote, in his 1918 book *Mysticism and Logic*, that:

> 'Mathematics, rightly viewed, possesses not only truth, but supreme beauty – a beauty cold and austere, like that of sculpture.'

Another famous British mathematician, G. H. Hardy, wrote in his book *A Mathematician's Apology* (1940):

> 'The mathematician's patterns, like the painter's or the poet's, must be *beautiful*, the ideas, like the colours or the words, must fit together in a harmonious way. Beauty is the first test; there is no permanent place in the world for ugly mathematics ... It may be very hard to *define* mathematical beauty, but that is just as true of beauty of any kind – we may not know quite what we mean by a beautiful poem, but that does not prevent us from recognising one when we read it.'

Both writers here were thinking of a highly abstract form of beauty, an inner beauty known to all professional mathematicians, but which for the great majority of us must remain forever unseen, and very likely not even dreamt of. It is a beauty of logical form and structure, of elegance of proof, a beauty which can be glimpsed only after a long and arduous apprenticeship has been served.

Or at least such was the case until the early 1980s, when the development of the electronic computer, and in particular its graphics facilities, gave rise to some new mathematical developments that changed everything. *Chaotic dynamics* is one of several names which have been given to one new area of mathematics that the computer has opened up. Though some of the mathematics involved in this subject is as hard and as abstract as any other examples of the mathematician's art, the essential beauty of the resulting structures can be displayed on a computer screen for all to see, professional and layman alike. Hard copies of computer-graphics displays formed the core of an exhibition organized by the German Goethe Institute, which began to tour the world in 1985, finding its way into both university mathematics departments and public art galleries alike. The film industry

Figure 8.   Fractal art – a view into Mandelbrot's world.

too was not slow to appreciate the potential of the new mathematics, and increasingly ideas from *complex dynamics* (to use another name for the same field) are being used in the creation of graphics for science-fiction films.

Figure 8 shows just one of a great many pictorial representations of the kinds of structure which are commonplace in this new field. (Many of them may be produced in colour to highlight patterns not visible in a black-and-white representation.) Remarkable though it may seem, the complexity in Figure 8 results from some quite simple mathematics (though a detailed *analysis* can involve very advanced methods). That mathematics is explained in this chapter.

How Long is the Coastline of Britain?

That was the question asked in an epoch-making article of the same title published in the magazine *Science* in 1967. The author was Benoit Mandelbrot, a brilliant French mathematician working at the IBM Thomas J. Watson Research Center at Yorktown Heights, New York. On the face of it the question seems innocuous enough, and you might expect that a pretty good answer could be obtained either with the aid of a map or by aerial reconnaissance. The only trouble is that no matter how carefully you do it, you will not get the right answer. And for a very good reason: *there is no right answer*! Mandelbrot arrived at this startling conclusion by reasoning as follows.

Suppose you make your measurement by flying around the coastline in a jet airliner at an altitude of 10 000 metres, taking photographs of the coast all the time, and then, using the appropriate scaling factor, calculating the total length as indicated by the vast collection of photographs you will have accumulated. How accurate is this answer? Not very. From 10 000 metres you will be unable to distinguish a great many small bays and promontories. (Let's assume that your camera is a good but otherwise quite ordinary model.) If you were to repeat the measurement from a small aeroplane flying at a height of 500 metres, so much extra detail will be visible that the answer you get will be significantly greater than your previous one. What on the first photograph appeared as a smooth stretch of coastline will now be found to consist of numerous little inlets, bays, and promontories.

Now suppose you set off on foot to measure the coastline using a pair of dividers set at a separation of, say, 1 metre. Then features of the coastline not visible from the air will result in an answer which is greater still. If the measurement is repeated with the dividers set at 10 centimetres the result will be even greater. And so on. Each time you make your scale of measurement finer, more detail of the coastline shows up and your answer gets larger. Soon you are measuring around pebbles, then around grains of sand, then molecules, and so on. And all the while your answer keeps on growing.

Of course, in the physical world this process of taking finer and finer measurements must come to an end eventually. Human limitations would probably bring you to a stop with the 1 metre dividers, whilst the physicist might argue that the procedure has a *theoretical* limit at the atomic level. But from the idealized viewpoint of the mathematician the process of making finer and finer measurements may be continued indefinitely. Since this means that the corresponding sequence of measurements increases indefinitely, it follows that there is no *mathematically precise* notion of the length of the coastline, only arbitrary choices – choices which are not even approximations to some 'real' answer.

An idealized, mathematical analogue of Mandelbrot's elusive coastline is provided by a geometrical figure first considered by H. von Koch in 1904, which we shall call *Koch's island*. Figure 9(i) shows Koch's island as seen from a rocket in outer space. From this distance it looks just like an equilateral triangle. As the rocket approaches Earth, it becomes clear that each of the three straight edges actually contains a central, triangular promontory, forming an equilateral triangle occupying the middle third of the line (Figure 9(ii)). If the perimeter length in Figure 9(i) is 3 units, then that in

Figure 9.   Construction of Koch's island.

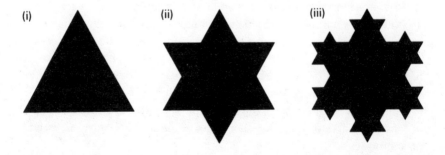

(i)   (ii)   (iii)

Figure 10. Koch's island taking shape.

Figure 9(ii) will be $3 \times (4/3)$ units. Coming in closer still, you see that each of the twelve straight edges you saw before likewise contains a promontory in the shape of an equilateral triangle occupying the middle third (Figure 9(iii)). The perimeter length now is $3 \times (4/3) \times (4/3)$ units. Figure 10 shows the island as seen from much closer, after several more levels of detail have unfolded, and gives some indication of the actual (?) shape of Koch's island.

To the mathematician, the nice feature of Koch's example is the regularity with which successive levels of detail appear. At each stage, the middle third of every straight-line segment of the coastline is replaced by two straight-line segments, each equal in length to that third, as shown in Figure 11.

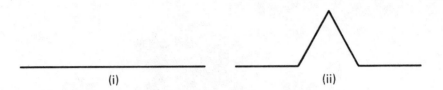

Figure 11.   Generation of the Koch coastline.

As you might surmise from an examination of Figures 9 and 10, Koch's island does have a (mathematically) well-defined shape, which Figure 10 approximates quite well as far as the human eye can distinguish. The mathematically precise *coastline* of Koch's island is the 'curve' which is the limit of the infinite sequence of approximations to it, of which Figure 9 gives the first three. At this point the mathematics takes over from the human cartographer. *Mathematically*, this limit curve is precisely determined, and like any other curve will consist of an infinitude of points strung together to form a 'line'. The process of arriving at the limit curve is analogous to arriving at the number 1/3 as the limit of the infinite sequence of decimals

$$0{\cdot}3,\ 0{\cdot}33,\ 0{\cdot}333,\ 0{\cdot}3333,\ 0{\cdot}33333,\ \dots .$$

Since Koch's island is a mathematically defined region of the plane, it will have a definite area. The actual numerical value of its area will depend on the units of measurement being used, of course, but it will certainly be *finite*. (It may be calculated as a limit of a sequence of numbers, much like the 1/3 example above; it is, in fact, exactly 1·6 times the area of the triangle in Figure 9(i).) What of the length of the coastline surrounding this finite area? Well, each successive stage of the Koch process increases the length of the 'coastline' by a factor of 4/3. By the time the *Koch curve* (as the limiting coastline is called) is reached, this 4/3 increase will have occurred infinitely often, and so the length of the Koch curve will be infinite.

How can a finite area have an infinite boundary? Figures 9 and 10 themselves provide the answer. The boundary curve twists from side to side along its entire length. For each of the finite approximations to the final curve, this twisting can be drawn in full provided you use a suitable scale (magnification), but for the actual Koch curve the twisting is infinite, and then something very strange occurs: a new dimension is entered.

New Dimensions

The curves that we usually meet in geometry are all *one-dimensional* : a creature constrained to live on, say, a straight line or a circle can travel in only one direction (if travelling backwards is regarded simply as negative forward movement). The usual geometrical surfaces such as planes or spheres are *two-dimensional* : there are two independent directions of travel, often referred to in terms of forwards/backwards and left/right. Solid objects are *three-dimensional*, allowing for three directions of motion. Railways provide an example of motion restricted to one dimension, ships can travel in two dimensions over the surface of the sea, and aircraft can move in three dimensions.

As far as human experience is concerned, there are only three dimensions to the universe we live in (though relativity theory regards time as a 'fourth dimension', and some current physical theories ascribe to the universe eleven dimensions, the three we are physically aware of plus a further eight which manifest themselves as the basic forces of nature, gravity, magnetism, and so on). But for the mathematician there is nothing special about three 'dimensions'. 'Spaces' of four or more dimensions may, and routinely are, considered. Though they cannot be realized by traditional geometry, such higher-dimensional spaces can be of real, practical use. (A case in point is the subject known as linear programming, considered in Chapter 11.) But notice that such 'higher dimensions' are still whole numbers.

Where does the Koch coastline fit into all this? Being a curve (in the mathematical sense, though with infinite twisting you could not hope to draw it), you might imagine it is one-dimensional, but this is not so. Although each of the approximations to the Koch curve that are obtained by means of the process described earlier is one-dimensional, the limiting curve is not. With the direction changing infinitely often, we are no longer in a familiar world – indeed, that use of the word 'direction' cannot really be justified. So we cannot hope to decide on the dimensionality of the Koch curve by speaking of 'direction of travel'. What we have to do is find another way of getting at the concept of dimension, one that does not depend on direction.

It makes sense to adopt an approach which is suited to the nature of the

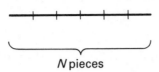

N pieces

Figure 12.   Self-similarity for a straight line.

Koch curve. The key feature is *self-similarity*: the parts are similar to the whole (only on a reduced scale).

Suppose we take a *D*-dimensional figure and divide it into *N* entirely similar parts. Then the *similarity ratio*, *r*, between the entire figure and a single part (i.e. the factor by which the whole exceeds the part in size) will be given by

$$r = \sqrt[D]{N}.$$

(Since the figure is *D*-dimensional and *r* has to be evaluated 'along a dimension', it is necessary to take the *D*th root of *N*.)

For example, suppose we take a straight line and split it into *N* equal pieces (see Figure 12). Then each piece is exactly $1/N$ of the length of the whole, so the similarity ratio will be *N*. This is precisely the value obtained from the above formula when you take $D = 1$.

Or how about taking a rectangle (so $D = 2$) and splitting it into *N* pieces by dividing it horizontally and vertically into *k* segments (see Figure 13)? Then the entire rectangle is split into exactly $N = k^2$ identical smaller replications of the whole, and the (linear) ratio *r* of the whole to any one of

Figure 13.   Self-similarity for a rectangle.

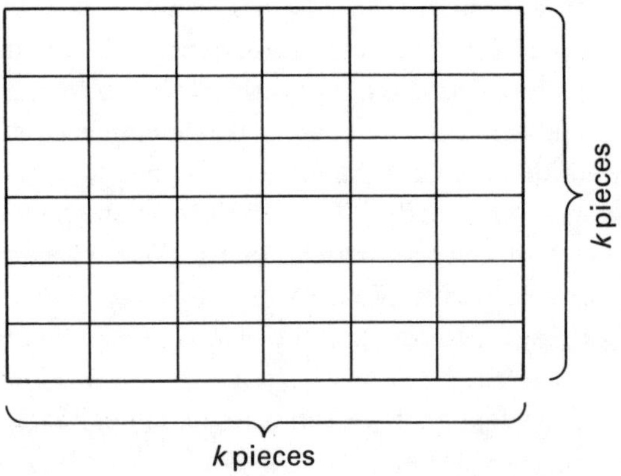

k pieces

k pieces

the parts is given by

$$r = \sqrt[D]{N} = \sqrt[2]{N} = \sqrt[2]{k^2} = k.$$

Again, this is exactly what you would expect.

In both these cases we seem to have been going round in circles, but this was because we were dealing with cases that are very familiar and not at all problematical. When we apply the same analysis to the Koch curve we arrive at an altogether more surprising conclusion. For this curve we do not know D, but the values of N and r are easily determined. All we need to do is look at the replication procedure that produces the curve. We first look at one section of coastline (see Figure 11(i)) – any one will do since they are all the same. In replication (see Figure 11(ii)) the single line is replaced by four lines (so $N = 4$), each one-third the length of the original line (so $r = 3$). Since this is true for any one piece of the coastline, it will be true for the whole Koch curve. So, according to the formula determined above,

$$3 = \sqrt[D]{4}.$$

So what is D? Certainly not a whole number. The only way to determine its value is by using logarithms. If you take logarithms of both sides of the above equation you get

$$\log 3 = D \log 4.$$

By reference to a book of logarithm tables (or by use of a calculator which can evaluate logarithms), D may be calculated; to four decimal places it is

$$D = 1\cdot2618.$$

So the Koch curve is a mathematical entity whose dimension is *fractional*.

It is not just 'curves' that can have fractional dimensions. Equally bizarre 'surfaces' and 'solids' may also be constructed using self-replication processes. For instance, by starting with a cube and successively removing 'middles' you arrive eventually (i.e. after infinitely many repetitions) at an object known as the *Sierpinski sponge* ($D = 2\cdot7268$), whose construction is shown in Figure 14. This incredible object has zero volume enclosed by an infinite 'surface'. Each external face of the sponge is known as a *Sierpinski carpet*, and has zero area surrounded by an infinite boundary. The dimension of the Sierpinski carpet is $D = 1\cdot2618$, the same as for the Koch curve. You should be able to confirm both values of D associated with the sponge

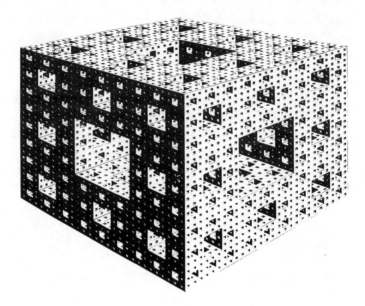

Figure 14.   The Sierpinski sponge taking shape.

by looking at Figure 14 and using the formula

$$r = \sqrt[D]{N},$$

or, taking logarithms,

$$D = \log N / \log r.$$

Figures having a fractional dimension were given the name *fractals* by Mandelbrot in 1977. *Fractal geometry* is the study of such objects.

The remainder of this chapter is also concerned with fractals, but ones of a different sort from the Koch curve and the Sierpinski sponge. These two fractals are highly regular. The self-replication process is the same at every level, and 'zooming in' on a particular part of the figure to see more detail produces no surprises – just more of the same, *ad infinitum*. Since 1980 computers have been used to examine fractals where the self-replication is continuously changing (though, as will become clear, it can often still be called 'self-replication'). With such figures, zooming in can produce totally unexpected results, of which Figure 8 is but one example. The examination of such fractals is a subject which is part mathematics, part experimentation (with the computer as the experimental tool), and it leads the researcher into a fascinating and often extremely beautiful new world. Like many other new worlds, its discovery was governed in part by chance.

Discovering a New World

By 1978, Benoit Mandelbrot's work on fractals was very well developed. The previous year had seen the publication of his book *Fractals: Form, Chance, and Dimension*, in which it was shown how many everyday phenomena in physics, biology, and mathematics give rise to fractals. All the fractals he had considered had been, like the Koch curve, self-similar. They gave rise to some interesting mathematics and occasionally startling conclusions, as well as some attractive and highly symmetrical figures (many of which are illustrated in Mandelbrot's book), but there was a built-in predictability in each of the examples that is not present in the real-life fractals of which they were mathematical counterparts. (For instance, the coastline of Britain exhibits a fractal behaviour far less ordered than Koch's coastline.) This extreme orderliness and predictability arises because the fractals considered were self-similar in the sense of scaling and translation (in mathematical language they were *invariant under linear transformations*). Working with Mark Laff at IBM in 1978/79, Mandelbrot began to investigate fractals which are invariant under *non-linear transformations* (where instead of a simple scaling, much more complicated manipulations are allowed, involving squaring, cubing, and so on). In such a case the only way to get any idea of what the corresponding fractal looks like is to get a computer to start generating it. Indeed, during the early part of this century, work on the same notions by Gaston Julia and Pierre Fatou in France had come to a halt, partly because there was no way of picturing the objects being considered. (Mandelbrot had been aware of this work from his student days at the École Polytechnique in Paris, where Julia had been one of his teachers.)

By the end of 1979, Mandelbrot had come to the conclusion that it was worth investigating – using the computer – the behaviour of the particular function $x^2 + c$, where both the variable x and the constant parameter c are complex numbers. (Precisely what 'behaviour' was being considered will be explained later, but suffice it for now to say that it is possible to use computers to draw diagrams relating this behaviour to varying values of the parameter c.)

Ironically enough, Mandelbrot was not at IBM for what was to be the crucial year, 1979/80, but was visiting Harvard University, and so did not have daily access to the famed 'unlimited computing facilities' of IBM at the time when his work most needed them. But in the basement of the Science Center at Harvard he found a newly delivered Vax supermini computer, to which were attached a rather old Tektronix visual display unit to view the output and a Versatec printer which could provide hard copies. A Harvard teaching assistant called Peter Moldave volunteered his services as an (unpaid) programmer for the project. And so the work went ahead.

The first picture they obtained was a crude version of the beetle-like double blob shown (in much greater detail) in Figure 20. This they had been expecting – the theory predicted it. More perplexing were a number of smaller blobs away from the main picture. Closer examination of these blobs revealed that they were smaller versions of the main beetle! It looked as though the familiar fractal reproductive behaviour was once more manifesting itself. By performing more accurate computations better pictures were obtained, showing more detail, until suddenly the pictures began to look increasingly messy. Perhaps their ancient graphics equipment was faulty? To make sure, Mandelbrot took the program to run it on an IBM mainframe computer at his home base at Yorktown Heights. Not only did the mess fail to disappear, but a better-quality picture showed that there was an underlying pattern to it. Homing in to take an even closer look, Mandelbrot and Moldave found that some of the small dust-like blobs were not smaller versions of the beetle, as they had supposed, but instead were beautiful and intricate patterns – spirals, families of figures like sea-horses, and the like (see Figures 8 and 23.) Mandelbrot had glimpsed his new world.

Order and Chaos

Order and chaos. Throughout history and throughout the universe the two compete for supremacy. Often only a knife edge separates them: a small change in the pressure can turn the ordered flow of water from a tap into a highly complex chaos of vortices; ordered animal populations (including human ones) can be transformed with frightening ease into uncontrollable anarchies. In the other direction order can emerge from chaos, as witnessed by the evolution of life – and ultimately mankind – from

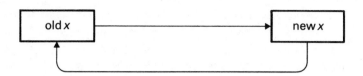

Figure 15. Feedback mechanism changing the value of *x*.

the formal chaos of the universe. As we shall see, the transition from order to chaos and the subsequent emergence of order from within that chaos appears in dramatic form with the study of simple *feedback loops*.

The essential feature of the feedback mechanism is that there is some quantity, call it *x*, that is changing, either over time (as in the example below) or else with respect to some other variable, in such a way that the value of *x* at any instant depends in a regular way upon its value at the previous instant (see Figure 15). Processes of this kind permeate all the exact sciences and most, if not all, of the inexact sciences. Much of modern mathematics was developed in order to deal with such processes. For example, it was in order to handle the case where the increment between the old *x* and the new *x* is infinitesimal that the many techniques for handling differential equations were developed.

To study a feedback process mathematically, the rule for generating the new value of *x* from the previous value is taken to be a mathematical function, $f(x)$. Then, starting with an initial *seed* value, x_0, of *x*, successive values x_1, x_2, x_3, ... are generated according to the rule illustrated in Figure 16. There need be no restriction on the function $f(x)$, though the resulting feedback process will not be very interesting unless the chosen $f(x)$ is something other than a *linear* function, one having the form

$$f(x) = ax + b$$

for some constants *a* and *b*. We shall be particularly concerned with the

Figure 16. Generation of successive values of *x* by feedback.

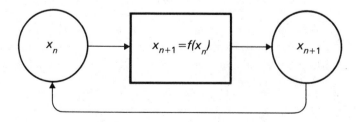

case where $f(x)$ involves a parameter. The choice of that parameter can have a dramatic effect on the behaviour of the resulting feedback process.

It is usual to regard the operation of a feedback loop as a *dynamical system*, which *sends* an initial point x_0 successively into the points x_1, x_2, x_3, ... The sequence of points to which x_0 is sent may be called the *path* or *orbit* of x_0. If this path is ordered we may speak of *ordered dynamics*; if it is not, it may be described as *chaotic dynamics*. This nomenclature alone should indicate how this study relates to many everyday phenomena.

As an example, consider the growth of a population over a number of years. Suppose the initial size of the population is x_0, and let x_n be the population after n years. The growth rate during the $(n + 1)$th year is then

$$r = \frac{x_{n+1} - x_n}{x_n}.$$

If the growth rate is constant from year to year, this equation will be valid for every value of n, and may be rearranged to give the (linear) *dynamical law*

$$x_{n+1} = f(x_n) = (1 + r)x_n.$$

After n years, the population will be

$$x_n = (1 + r)^n x_0$$

(an expression obtained by working backwards from $x_n = (1 + r)x_{n-1}$, $x_{n-1} = (1 + r)x_{n-2}$, and so on, down to $x_1 = (1 + r)x_0$). This is an example of what is known as *exponential growth*, and is typical (up to a point) of many real-life phenomena besides population growth. As should be clear from what we have seen in Chapter 1, if continued without check for a number of years, such a growth mechanism will lead to enormous populations. What happens in practice is that such growth will occur over only a limited period, after which a limit is reached. In 1845, P. F. Verhulst formulated a growth law which takes account of the existence of a maximum possible population size, say X. Verhulst's law says that the growth rate will drop from r to 0 as the population approaches X. A simple way to represent this mathematically is to replace the constant growth rate r by the variable growth rate $r - cx_n$, where c is some constant. Since the population growth should become zero when $x_n = X$, the value of the constant c will have to be r/X. With this value, the dynamical law for Verhulst's process is thus

$$x_{n+1} = f(x_n) = (1 + r - cx_n)x_n = (1 + r)x_n - cx_n^2.$$

Once the value X has been reached, the population will remain constant:

$$f(X) = X.$$

If the population is smaller it will grow larger. If it is larger it will decrease. If you try it out (either by hand or using a computer) you will see that the Verhulst process will give a population evolution which eventually becomes stable at size X, regardless of the initial size. At least, that is what happens provided r is less than 2 (i.e. a 200% growth rate) – a restriction that certainly applies to the growth of human populations. But, as was observed by the meteorologist E. N. Lorenz in 1963, for larger values of r the Verhulst law describes certain aspects of turbulent flow, and there are other applications in laser physics, in hydrodynamics, and in the theory of chemical reactions, so the behaviour of the Verhulst process for values of r greater than 2 is not without interest. And it turns out that this is where the really fascinating results are found.

Putting $c = r/X$, the rule above becomes

$$x_{n+1} = (1 + r)x_n - (r/X)x_n^2.$$

By altering the units of measurement appropriately we may assume that $X = 1$, so the rule simplifies even further to

$$x_{n+1} = (1 + r)x_n - rx_n^2 = x_n + rx_n(1 - x_n).$$

If you have access to a computer you can easily perform some experiments to see how the Verhulst process evolves for different values of r in this last formulation of the equation, starting with the seed value (say) $x_0 = 0\cdot1$ in each case. (Your program should read in the chosen value of r, set $x = 0\cdot1$, iterate the operation

$$x = x + r * x * (1 - x)$$

500 times to give the process time to settle down, and then calculate and print out the next 20 or so values of x.) For values of r less than 2 the process quickly settles down to the equilibrium value of $x = 1$. For r just above 2 the process settles into a regular oscillation between two values ($r = 2\cdot1$ gives the values $0\cdot82$ and $1\cdot13$). This behaviour continues for all choices of r up to $r = 2\cdot5$, when there is a repeated cycling through four points ($0\cdot54$, $1\cdot16$, $0\cdot70$, $1\cdot23$). This continues up to $r = 2\cdot55$, when a cycle

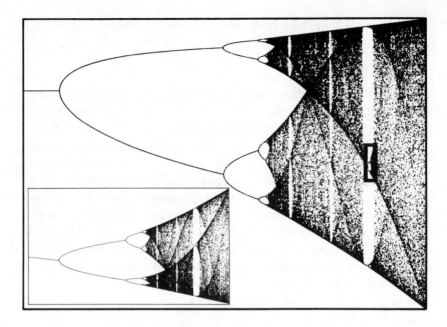

Figure 17. The Verhulst process ($1 \cdot 9 < r < 3 \cdot 0$), with a blow-up of the area indicated, showing self-replication. Values of r from $1 \cdot 9$ to $3 \cdot 0$ are plotted along the horizontal axis. For each value of r, 120 successive values of x are plotted against the vertical axis after an initial run of 5000 iterations to allow the process to settle down. For values of r less than 2 only one x-value arises. For r between 2 and $2 \cdot 5$ there are two values, for r between $2 \cdot 5$ and $2 \cdot 55$ four, then eight values as far as $2 \cdot 565$. This doubling up continues ever more rapidly until $r = 2 \cdot 57$, where chaos sets in. But within this chaos further order begins to emerge, including self-replication.

through eight values begins. For $r = 2 \cdot 565$ it doubles up yet again to sixteen values through which the process eventually cycles indefinitely, and so on and so on, continuing to double ever more frequently until at $r = 2 \cdot 57$ the doubling-up effect has occurred infinitely often. At this point the behaviour of the dynamical system becomes chaotic, jumping around all over the place with no apparent pattern.

The various cycles to which the process converges for values of r less than $2 \cdot 57$ are known as *attractors*. So for r less than 2 the attractor consists of one point, the fixed point $x = 1$; for r between 2 and $2 \cdot 5$ the attractor is an oscillating pair of values; for r between $2 \cdot 5$ and $2 \cdot 55$ it is a cycle of four points; and so on.

A clearer picture of what is going on can be obtained by drawing a graph to relate the behaviour of the process (after the initial settling-down period) to the appropriate value of r. The main part of Figure 17 shows the result of taking values of r from $1\cdot9$ to $3\cdot0$, measured along the horizontal axis, and plotting 120 successive values of x after an initial settling-down period of 5000 iterations.

A careful analysis of the chaotic region above $r = 2\cdot57$ reveals that beneath this chaos lies a considerable amount of order. For instance, near $r = 3\cdot0$ there is only one chaotic band, at $r = 2\cdot679$ this splits into two chaotic bands, at $r = 2\cdot593$ into four, then into eight, sixteen, and so on, doubling up each time until at $r = 2\cdot57$ this doubling up has occurred infinitely many times, the entire process replicating the behaviour of the dynamical system itself. In fact there is a *universal constant* related not only to both doubling-up processes encountered so far, but also to all other examples of this phenomenon. It is called the *Feigenbaum number*, whose value to ten decimal places is

$$4\cdot669\ 201\ 660\ 9.$$

Much more apparent of course is the appearance of bands within the chaotic region, where order seems to reign (briefly) once more. For instance, near $r = 2\cdot83$ the chaos suddenly gives way to a three-cycle attractor. And in the region around its middle point, what do you discover but a tiny replica of the entire Verhulst diagram, complete with its own ordered bands amidst the chaos. (An enlargement of this region is shown in the inset to Figure 17, 'stretched' in the horizontal direction.) Fractal behaviour once more! But this is only the beginning.

Julia Sets

Mandelbrot's work in 1980, mentioned earlier, followed on from work (mainly by P. J. Myrberg in the 1960s) on the Verhulst process. The main difference in Mandelbrot's approach was to allow the variable and the constant parameter to be complex numbers rather than just real numbers. So instead of the process sending numbers to numbers (on the real line), it sends points to points (in the two-dimensional complex plane, or Argand diagram).

To make things slightly easier, instead of taking the function

$$f(x) = x + rx(1 - x) = -rx^2 + (1 + r)x,$$

as above, Mandelbrot used the slightly simpler formula

$$f(x) = x^2 + c,$$

and this is the function we shall consider here.

Suppose you start with some seed value x_0 (a complex number), then see what happens when you iterate the function f to generate a sequence of points $x_0, x_1, x_2, \ldots$ according to the rule

$$x_{n+1} = f(x_n).$$

The results obtained for the Verhulst process suggest that the choice of the constant parameter c will have a significant effect. Suppose we look at the simplest case, when $c = 0$. The dynamical law is then just

$$x_{n+1} = x_n^2.$$

There are then three possible outcomes, depending on the choice of x_0. Firstly, if x_0 is less than a distance 1 unit from 0 (the origin), then the numbers in the sequence become smaller and smaller (i.e. closer and closer to 0), which is to say that 0 is an attractor for the system. Secondly, if x_0 is a greater distance than 1 unit from 0, then the numbers in the sequence get larger and larger, in which case we say that infinity is an attractor (though as infinity is not a *point* in the complex plane, this particular use of the word 'attractor' is purely a stylized one). The remaining possibility is when x_0 is exactly 1 unit from the origin (i.e. x_0 lies on the unit circle centred at 0). In this case the sequence never leaves the unit circle. Thus the unit circle is the *boundary* between the two domains of attraction, the one governed by 0, the other by infinity.

In so far as it divides the complex plane into two distinct regions of attraction separated by a bounding curve, this example is typical of all the cases considered by Mandelbrot (and others). But what Mandelbrot discovered was that for non-zero values of the parameter c, not only can the non-infinite attractor consist of more than one point, but also the boundary between the domains of the two attractors can be incredibly complex and extremely beautiful.

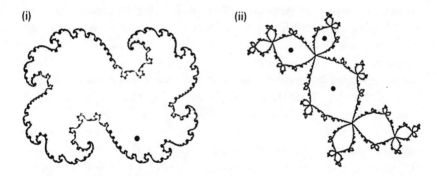

Figure 18.   Julia sets with their attractors (see the text for details).

For $c = 0\cdot31 + 0\cdot04i$, for example, the non-infinite attractor is a single point, but the boundary between its domain and that governed by infinity is not a perfect circle, but rather the attractively deformed circle shown in Figure 18(i). It is a fractal deformation: if you move in to take a closer look at any one part of the boundary (using a computer as your 'microscope'), you find the familiar, endlessly repeating self-similarity that is characteristic of fractal curves.

Figure 19.   Julia sets from the boundary of the Mandelbrot set (see the text for details).

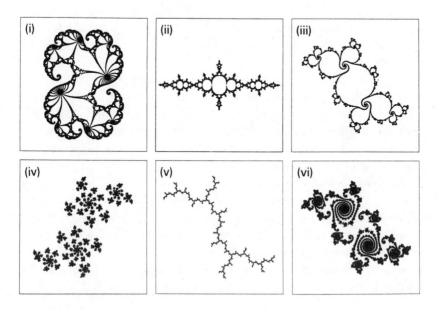

Though it was only with the advent of the computer that it became possible to examine such figures, it had been proved by Julia and Fatou that any piece of the boundary, no matter how small, will contain all the information required to determine the whole curve (in that the entire boundary can be generated by repeatedly subjecting the piece to the transformation which determines the system, in this case $f(x) = x^2 + c$). In honour of Julia, such boundary sets are nowadays known as *Julia sets*.

Figure 18(ii) shows a Julia set associated with a dynamical process having a non-infinite attractor consisting of a cycle through three points. The dynamical law here is $f(x) = x^2 + c$, with $c = -0.12 + 0.74i$. Figure 19 shows other examples of Julia sets that arise from the law $f(x) = x^2 + c$, including some extreme cases where the regions degenerate into 'dust' or 'dendrites' (see later for details).

The diversity of structure exhibited by the Julia sets for differing choices of the parameter highlights just how critical is the choice of c. A natural question to ask is whether there is any discernible pattern to the values of c for which the corresponding dynamical system and its Julia sets has a particular form. Investigation of this question led Mandelbrot to his 1980 discovery of the region (subset) of the complex plane that now bears his name: the Mandelbrot set.

The Mandelbrot Set

The black beetle-shaped blob shown in Figure 20 is known as the *Mandelbrot set*. Since its discovery it has been shown that this set is intimately connected with the behaviour of all dynamical processes, not just the single example considered here. As such it occupies a special, fundamental place in mathematics, along with other special shapes such as the circle and the regular polygons.

As should be clear from a glance at Figures 18 and 19, a complex dynamical process either divides the complex plane into one or more interior regions and a single exterior region stretching to infinity (Figure 18(i), (ii), Figure 19(i), (ii), (iii)), or it causes the Julia set to degenerate into a set which borders no inner region (Figure 19(iv), (v), (vi)). The exact behaviour depends upon the location of the parameter c relative to the Mandelbrot set. Continuing with the example of the process $f(x) = x^2 + c$,

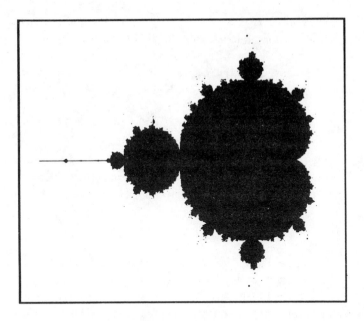

Figure 20. The Mandelbrot set ($-2 \cdot 25 <$ Re $c < 0 \cdot 75$, $-1 \cdot 5 <$ Im $c < 1 \cdot 5$) (see the text for details).

we shall first of all consider those cases where there is a non-degenerate Julia set. In this case there is an attractor other than infinity.

If c is chosen from within the main body of the Mandelbrot set, then the associated dynamical system has a non-infinite attractor consisting of a single point (a fixed point, satisfying $f(x) = x$). The Julia set in this case is a fractally deformed circle, as in Figure 18(i). (The constant c here is located near the right-hand edge of the cardioid-shaped main body of the Mandelbrot set.)

If, on the other hand, c is chosen from within one of the buds attached to the main body of the Mandelbrot set, then the Julia set consists of infinitely many fractally deformed circles surrounding the points of a periodic-cycle attractor, and the points which are eventually sent to them. For Figure 18(ii), for instance, c is chosen from the centre of the large bulb at the top of the Mandelbrot set. The three points indicated form the 3-cycle of the non-infinite attractor for the system. A point chosen from within any of the three regions containing this attractor will move directly in towards that 3-cycle; points from within the other regions will be attracted towards a local 'attractor' which is sent into the 3-cycle.

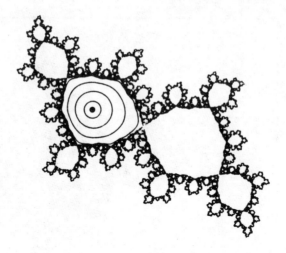

Figure 21.   Siegel disc (see the text for details).

If c is the germination point of a bud on the Mandelbrot set, the Julia set turns out to have tendrils that reach towards a marginally stable attractor, as in Figure 19(i), which eventually settles into a 20-cycle ($c = 0 \cdot 273\,34 + 0 \cdot 007\,42\text{i}$), or Figure 19(ii) which has a 4-cycle ($c = -1 \cdot 25$).

Finally, if c is any other boundary point of the Mandelbrot set, the Julia set turns out to be what is known as a *Siegel disc*, an example of which is shown in Figure 21 ($c = -0 \cdot 390\,54 - 0 \cdot 586\,79\text{i}$). Here there is a fixed point surrounded by *invariant circles*. What happens in this case is that a point within the region bounded by the Julia set will gradually make its way in towards the disc containing the fixed point, whereupon it will orbit for ever around the fixed point on its invariant circle.

The four types of Julia set described above are the only ones possible for the process $f(x) = x^2 + c$. (In 1983, Dennis Sullivan showed that there is one further type of non-degenerate Julia set that may arise in other complex dynamical systems – the Herman ring.)

So much for the non-degenerate Julia sets. But what of the others, as in Figure 19(iv), (v), (vi)? Large-scale pictures of the Mandelbrot set show that it is surrounded by hair-like, branching antennae. If c is chosen from one of these antennae, a similar-shaped Julia set is obtained. Figure 19(v) shows the example for $c = \text{i}$. The behaviour here is that the only attractor is infinity, and all points are sent there except for the ones actually on the hair-like Julia set.

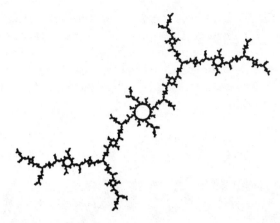

Figure 22. A Julia set from an outlining Mandelbrot sprout (see the text for details).

Figure 20 is not sufficiently detailed to show the antennae themselves, but the location of some of them can be discerned from some isolated dots that lie on their path. Dots? A close examination (by a computer blow-up) would reveal that they are nothing but tiny replicas of the Mandelbrot set itself! They even have tiny antennae around them, along which may be found … and so on, *ad infinitum*. (The gaps in the chaotic region of the Verhulst process (see Figure 17) correspond to the positions of these 'sprouts' relative to the real axis.) If c is chosen from one of these sprouts, the corresponding Julia set is a combination of a dendrite and infinitely many copies of the Julia set from the corresponding value of c in the main Mandelbrot set (see Figure 22).

The only remaining possibility is to choose c from outside the Mandelbrot set (with all its attachments) altogether. Here infinity is the only attractor, and the Julia set dissolves into isolated points called *Fatou dust*. This dust gets thinner and thinner the further c is from the Mandelbrot set. If c is chosen from a point near the boundary of the Mandelbrot set, the dust is sufficiently thick to create fascinating patterns, as in Figure 19(iv), (vi). (For Figure 19(vi), c is close to the value that produces Figure 19(iii), and there is a noticeable similarity between the two Julia sets.) Such patterns in the dust are always fractal-like (i.e. self-similar), with chaotic dynamics.

Not surprisingly, with the boundary of the Mandelbrot set playing such a critical role in the dynamics of the associated systems, this boundary is itself of great interest. As you are probably expecting by now, this boundary region itself turns out to have a complicated, fractal surface. Figure 23 provides just one glimpse into this incredible world, a world where every single point of territory is contested, a world which is accessible only via the computer – with the degree of detail that can be discerned depending on the power of the computer. If any branch of mathematics is truly of this computer age, surely this must be it.

Figure 23.   A journey into the border region of the Mandelbrot set.

(i)

(ii)

(iii)

Suggested Further Reading

A true appreciation of the world just described can really only be obtained with the aid of colour representations, where colours can be used to provide a kind of relief or contour map reflecting the dynamics of the system. An excellent compilation of both colour and black-and-white pictures (with explanations) is provided by the book *The Beauty of Fractals*, by H. O. Peitgen and P. H. Richter (Springer-Verlag, 1986). This book is highly recommended.

For a fairly up-to-date account of fractal geometry and its many applications, see *The Fractal Geometry of Nature*, by Benoit B. Mandelbrot (W. H. Freeman, 1982).

5 Simple Groups

The Enormous Theorem

Some time during the summer of 1980, Ohio State University mathematician Ronald Solomon put down his pen after solving a technical problem in algebra, and with that one simple action came the end of a quest that had begun in the 1940s and which had involved over a hundred mathematicians from the USA, Britain, Germany, Australia, Canada, and Japan. For what Solomon's result did was to fill in the last piece of an enormous and highly complex puzzle: the classification of the finite simple groups.*

The classification theorem is by far the biggest theorem mathematics has ever seen. The initial proof runs to nearly 15 000 pages spread across some 500 articles in mathematical journals, and over 100 mathematicians contributed to that proof. Along the way, discoveries were made that led to advances in the theory of computer algorithms, in mathematical logic, in geometry, and in number theory, and there has been speculation that there may even be applications in the formulation of a unified field theory in physics.

And yet, as with many deep results in mathematics, the story has very humble beginnings: in this case the familiar formula

$$x = \frac{-b \pm \sqrt{b^2 - 4ac}}{2a}$$

*All technical terms will be explained in due course.

for the roots of the quadratic equation

$$ax^2 + bx + c = 0,$$

and attempts to obtain similar solutions for equations of higher degree (i.e. involving powers of x greater than 2, such as the cubic equation below). (By a 'similar solution' is meant one which involves only the basic algebraic operations of addition, subtraction, multiplication, and division, as well as the extraction of roots. Such solutions are sometimes referred to as solutions by radicals.)

Évariste Galois

Examination of ancient tablets makes it clear that the Babylonian mathematicians of 1600 BC knew how to solve quadratic equations, though they possessed no algebraic notation to express their equations and solutions as we do nowadays. The solution (by radicals) of a cubic equation, i.e. one of the form

$$ax^3 + bx^2 + cx + d = 0,$$

was not discovered until the sixteenth century, when the Italian mathematicians Scipio de Ferro and Nicola Fontana independently found the method. Girolamo Cardano published Fontana's solution in his book *Ars Magna* in 1545. This volume also contained Ludovico Ferrari's method for solving a quartic equation (by reducing it to a cubic). But there the matter seemed to end. Despite the efforts of many mathematicians – including the great Swiss mathematician Leonhard Euler in the middle of the eighteenth century – no one was able to obtain a solution to the quintic equation

$$ax^5 + bx^4 + cx^3 + dx^2 + ex + f = 0.$$

Joseph Louis Lagrange in 1770 suggested that such a solution (i.e. one by radicals) might not be possible; in 1824 the Norwegian mathematician Niels Henrick Abel proved that this was indeed the case.

If there is no general method (i.e. no formula) for solving a quintic equation, it is natural to ask whether there is any way of deciding whether or not a *given* quintic equation can be solved (by radicals). Abel himself was wrestling with this problem when he died in 1829 at the age of 26. By which time, as it happens, the young man who was eventually to solve the problem was also working hard on it. But the remarkable results obtained by the young Évariste Galois were not to be recognized by the mathematical community until some eleven years after his death in a duel. And thereby hangs a tale, from which only the young man himself emerges with any credit.

Galois was born near Paris in October 1811. His interest in mathematics began when, at the age of 14, he was forced to repeat his third-year classes at the *lycée* after failing his examinations. Mathematics, he found, helped to stave off the boredom he felt for the rest of his schooling. Unfortunately, his growing passion for mathematics caused his schoolwork to deteriorate even further, and when he took the examination for entry into the prestigious École Polytechnique at the age of 15, he failed and was forced to go to the more ordinary École Normale. It was there in the following year that he produced his first paper on mathematics: a competent though unremarkable piece of work on continued fractions. A promising start, but soon after began a series of unfortunate mishaps that were eventually to lead to his complete abandonment of the subject he loved with such passion.

His next two papers (on polynomial equations) were rejected by the French Academy of Sciences. Worse, both manuscripts were unaccountably lost. Then, in July 1829, he failed once again to gain entry to the École Polytechnique, an occurrence which may have had a great deal to do with his answer to one particular question put to him by the examiner. When asked for an outline of the theory of 'arithmetic logarithms', Galois answered (with total accuracy but demonstrating an amazing lack of tact and understanding) that there were no *arithmetic* logarithms. Following this disappointment, early in 1830 Galois presented yet another paper to the Academy, this time in competition for the Grand Prize in Mathematics. The secretary, Fourier, took the manuscript home to read, but died before he had made his report, and the paper was never found. This third loss of work he had submitted, coupled with his repeated failure to gain entry into the Polytechnique, drove Galois to reject the academic community, and he became what would nowadays be called a student radical. Within the year he had been expelled from the school and forced to try to support himself through giving private tuition. Though in this he was not very successful, his mathematics continued to flourish, and it was during this period that he wrote what was to become his most famous paper, 'On the conditions of solubility of equations by radicals', submitted to the Academy in January 1831.

The submission of this paper was his final attempt to have his work recognized. When, by March, he had heard nothing from the Academy, he wrote to the president to find out what had happened to his paper. Receiving no reply to that letter, he finally gave up. He would do no more mathematics. Instead he joined the National Guard (a Republican organization). But here he turned out to have no more luck than in mathematics. Soon after he had joined, the Guard was disbanded following conspiracy charges. At a banquet held in protest on 9 May, Galois proposed a toast to the king with an open knife in his hand – a gesture which his companions not unnaturally interpreted as a threat to the king's life – and the following day he was arrested. At his trial he claimed that what he had actually said was 'to Louis Philippe, *if he turns traitor*', but that the uproar had drowned the last phrase. Whether or not this was true, he was acquitted and freed on 15 June.

On 4 July he finally learnt the fate of his paper to the Academy. Rejecting it as 'incomprehensible', the referee, Poisson, ended his report like this:

> 'We have made every effort to understand Galois's proof. His reasoning is not sufficiently clear, sufficiently developed, for us to judge its correctness, and we can give no idea of it in this report. The author announces that the proposition which is the special object of this memoir is part of a general theory susceptible of many applications. Perhaps it will transpire that the different parts of the theory are mutually clarifying, are easier to grasp together rather than in isolation. We would then suggest that the author should publish the whole of his work in order to form a definitive opinion. But in the state which the part he has submitted to the Academy now is, we cannot propose to give it approval.'

Connoisseurs of condescending rejections may feel that this report is a classic, and we have no idea whether or not its receipt had any influence on what Galois did next. But on 14 July he was arrested for appearing in public in the uniform of the now-disbanded National Guard, and was sentenced to six months' imprisonment.

Shortly after his release on parole he fell in love with a certain Mlle Stephanie D. (Her surname is not known. It did appear in one of Galois' manuscripts but had been heavily obliterated, possibly as a result of her subsequent rejection of him.) This was to lead to his early death. Somehow or other the *affaire* led to him being challenged to a duel. (Alexandre Dumas suggested that the duel was really a disguised assassination plot, with political motives.) On 29 May, the eve of the duel, Galois wrote a long letter to his friend Auguste Chevalier in which he outlined his mathematical theories, thereby providing the mathematical world with its only indication of the loss it was about to suffer. In the following day's duel (pistols at 25 paces) Galois was hit in the stomach and died twenty-four hours later.

And what of his rejected paper? On 4 July 1843, Joseph Liouville addressed the French Academy, opening with these words:

> 'I hope to interest the Academy in announcing that among the papers of Évariste Galois I have found a solution, as precise as it is profound, of this beautiful problem: whether or not it is soluble by radicals ... .'

The central concept which Galois had left to the world was to prove one of the most significant of all time, having applications in many fields of mathematics as well as in physics, chemistry, and branches of engineering. The concept is that of a *group*.

It is an entirely abstract concept. What makes it so powerful is the great number of examples of groups, examples of often quite different natures. Because of the versatility of the group concept, there are various ways of introducing it. The one chosen here uses symmetry properties of geometric figures in the plane, but this is purely because it provides easily visualized examples. Later on in this chapter we shall meet other types of groups, every bit as genuine as the symmetry groups we now turn to.

Symmetry

Consider the isosceles triangle illustrated in Figure 24. In everyday parlance this geometrical figure is 'symmetrical' about the vertical line indicated. What we mean by the statement that the triangle ABC is *symmetrical* is that the part of the triangle to the left of the vertical line (namely the smaller triangle ABD) is the mirror image of that part to the right (namely ACD) with respect to an imaginary mirror placed along the vertical line AD, perpendicular to the plane. If we were to swap over (or *reflect*) the two halves of the figure, the result would be an entirely similar triangle in exactly the same position, but with the lines AB and AC interchanged and the line BC reversed.

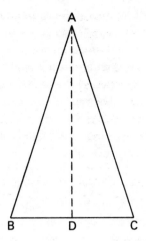

Figure 24. Symmetry of an isosceles triangle.

Figure 25. Reflections in an axis. In each case the figure shown in **bold** is reflected in the axis indicated by the dashed line, producing the lighter image. In (iv) the image coincides with the original figure.

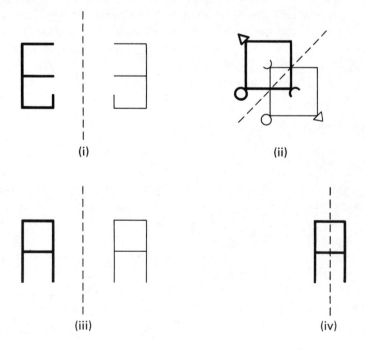

In general terms, for any geometric figure S in the plane and any line *l* in the plane, the *reflection* of S *in the axis l* is the action of moving every point of S to its mirror image in *l* – to the point an equal distance from *l* along the line drawn through the point and perpendicular to *l*. Notice that it is the *action* of transforming the figure that is referred to as the *reflection*, not the result of that action. (For reasons that will become clear we concentrate on actions rather than their results.) The figure produced by applying a reflection to a figure S is called the *image* of S under that reflection. Figure 25 shows some examples of (the results of) reflections.

Using the notion of a reflection, the mathematician says that a figure S in the plane is *symmetrical* about an *axis of symmetry l* if the reflection of S in *l* results in an image which occupies exactly the same position in the plane as does S. (Figure 25(iv) shows an example of symmetry about an axis.) Symmetry about an axis is sometimes referred to as *bilateral symmetry*.

Bilateral symmetry is pretty well what is meant by the everyday use of the word 'symmetry' (for planar figures), but for the mathematician there is another kind of symmetry. This is illustrated in Figure 26. If the shape shown is rotated through an angle of 120° (in either direction) about the central point, it will end up occupying exactly the same position in the plane. This is an example of *rotational symmetry*. (Note that it is rotation *about a point* that we are considering. Rotating the triangle in Figure 24 through 180° about the line AD as an axis also brings the figure to the same position, but the result is the same as reflection in AD.)

Figure 26.   Rotational symmetry.

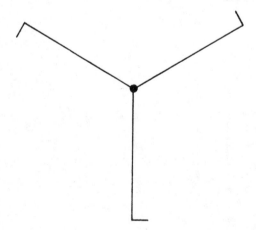

Groups

Consider the isosceles triangle in Figure 24 again. How many symmetries does it have? That is to say, what are the reflections (in axes) and rotations (about points) which send the triangle to an image occupying exactly the same position as the original triangle? Well, there is reflection about the vertical line AD. Let us denote this reflection by the letter r. Are there any other symmetries? Obviously there are no other reflectional symmetries, but how about rotations? Certainly a rotation through $360°$ about any point will bring the figure back to its starting point, but this does not really count, for in this case the result would be no change at all (whereas in the case of the reflection r there is a genuine change in that the points B and C end up occupying different positions from where they started). So we discount such trivial examples – or at least we almost do. But just as it is useful to consider the number 0 (which results in no change when we are adding) and the number 1 (which has no effect when we multiply), so too it proves useful to include as a symmetry the *identity transformation, I,* which leaves every point of the plane unchanged. I may be regarded as a rotation through $0°$.

Figure 27. Successive reflections.

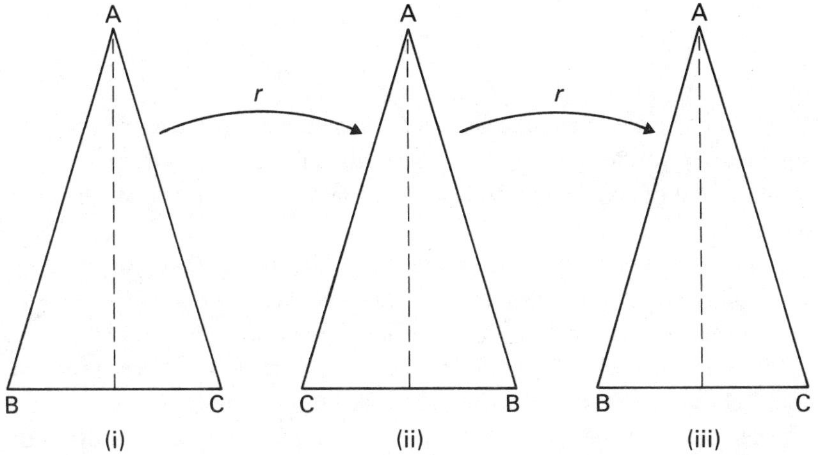

Suppose now we take the triangle ABC (see Figure 27(i)) and perform the reflection *r*, to produce the triangle ACB shown in Figure 27(ii). What happens when we perform *r* again, this time on ACB? Obviously we end up with the original configuration ABC again – Figure 27(iii). So the result of performing *r* twice in succession is the same as simply doing nothing (which we can express as the performance of the identity transformation *I*). We can express this symbolically by writing

$$r * r = I,$$

where the asterisk * means 'apply again'. (So if *a* and *b* were two symmetries, *a* * *b* would mean the operation consisting of first applying *a*, and then applying *b* to the result.) Using the same notation we may describe the (in this case trivial) effects of performing other sequences of symmetries, thus:

$$r * I = r,$$
$$I * r = r,$$
$$I * I = I.$$

These four identities may be summarized in a table:

Isosceles triangle:

*	I	r
I	I	r
r	r	I

To see the effect of the application of a symmetry *x* followed by another symmetry *y*, you look along the *x* row of the table until you come to the *y* column, and the entry you find is *x* * *y*, the result of the two symmetries combined.

Notice that it was implicit in the above discussion that the result of performing two symmetries in succession was itself a symmetry. This is indeed the case (you will see it is obvious if you think about it).

What happens when we perform the same analysis on the three-pronged shape shown in Figure 26? Here there are three symmetries: a rotation anticlockwise through 120° (call it *v*), a rotation anticlockwise through

240° (call it *w*), and the identity *I*, where everything is left fixed. ('What about clockwise rotations?' you might ask. Well, a clockwise rotation through 120° gives the same result as *w*, and a clockwise rotation through 240° is equivalent to *v*, so we really have included all possibilities.) Since two successive rotations through 120° have the same effect as one rotation through 240°, then obviously

$$v * v = w.$$

Similarly, two 240° rotations are equivalent to one 120° rotation, so

$$w * w = v.$$

The full table of successive symmetries is:

Tripod:

*	I	v	w
I	I	v	w
v	v	w	I
w	w	I	v

One more example. The equilateral triangle (see Figure 28) has six symmetries. There is the identity, *I*, anticlockwise rotations *v* and *w* through 120° and 240°, respectively, and reflections *x*, *y*, *z* in the lines X, Y, Z, respectively. (The lines X, Y, Z stay fixed when the triangle moves.) These symmetries combine as indicated by the following table:

Equilateral triangle:

*	I	v	w	x	y	z
I	I	v	w	x	y	z
v	v	w	I	z	x	y
w	w	I	v	y	z	x
x	x	y	z	I	v	w
y	y	z	x	w	I	v
z	z	x	y	v	w	I

If you wish to check these entries you could try cutting out an equilateral triangle from cardboard, marking the corners A, B, C, and placing it on a sheet of paper on which the lines X, Y, Z are drawn. Then you can physically perform the various rotations and reflections. (You will have to mark your triangle on both sides to allow for reflections.)

What about combinations of more than two symmetries? Well, there is no need to consider them, since the application of any number of symmetries is equivalent to a succession of pair combinations. For example, for the equilateral triangle just considered, $(x * y) * v$ is the same as $v * v$, which is just w, using the table twice. The parentheses were required in order to indicate the order in which the symmetries were to be combined: apply x followed by y, and then apply v to the result. The alternative grouping $x * (y * v)$ would mean: apply x and then apply $y * v$ to the result. If you did things the second way, what would you get? Well, $y * v$ is z, so $x * (y * v)$ is $x * z$, which is w, the same result as was obtained the other way. If you think about it for a moment you will realize that this is no accident. It is true for all combinations of symmetries: if a, b, c are symmetries of some figure, then

$$(a * b) * c \;=\; a * (b * c).$$

(This property is known as *associativity* of the operation $*$.)

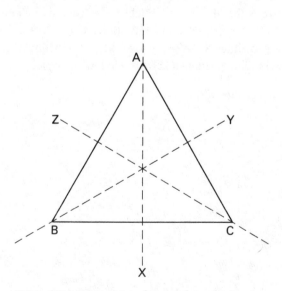

Figure 28. Symmetries of an equilateral triangle.

One final observation needs to be made before the definition of a 'group' can be given. It is obvious that if you take any symmetry and 'apply it backwards' then the result is another symmetry. (For a reflection there is no difference between the 'forward' and the 'backward' applications, of course. For rotations you simply reverse the direction of rotation.) The 'backward application' of a symmetry x is called the *inverse* of x, often denoted by the symbol x^{-1} (which you should read as either 'x to the minus one' or 'x-inverse'). For the isosceles triangle you get $r^{-1} = r$ (as for any reflection). For the tripod you get $v^{-1} = w$ and $w^{-1} = v$, and for the equilateral triangle $v^{-1} = w$, $w^{-1} = v$, $x^{-1} = x$, $y^{-1} = y$, and $z^{-1} = z$. In all cases $I^{-1} = I$. Do you notice anything about these results? If you check the tables in each case you will notice that it is always true that $a^{-1} * a = a * a^{-1} = I$. Once again this is no accident, and if you think about what is meant by the identity symmetry and the inverse of a symmetry, you will see why it happens.

By now you should have a vague feeling that there is something very familiar about all of the above – even if you have never thought about symmetries before. Doesn't it all look remarkably like ordinary multiplication of rational numbers (fractions), especially if you exclude zero (which has no inverse)? The product of any two non-zero rational numbers is another rational number, grouping does not matter for multiplication (i.e. $(ab)c = a(bc)$ for any numbers a, b, c), and every non-zero rational number x has an inverse rational number $x^{-1} (= 1/x)$ such that $xx^{-1} = x^{-1}x = 1$, where that number 1 has the property that multiplication by it causes no change. Or perhaps this was not quite what you were thinking of. Perhaps the example you had in mind was the whole numbers (integers) with addition: the sum of two integers is another integer, $(a + b) + c = a + (b + c)$ always holds, there is an identity number 0 which causes no change when you are adding, and every integer x has an inverse $(-x)$ such that $x + (-x) = (-x) + x = 0$. Or maybe you had yet another example in mind. There are, in fact, many possibilities, including the one Galois was thinking of when he analysed the problem of solving quintic equations. All these examples are special cases of Galois' general concept of a 'group'.

To the mathematician, a *group* consists of:

(1) a set G and

(2) an operation $*$ which to any pair of elements x and y of G assigns an element $x * y$, which also belongs to G.

The operation * is required to satisfy the following three conditions ('group axioms'):

(3) It is associative: for any x, y, z in G,

$$(x * y) * z \;=\; x * (y * z).$$

(4) There is an identity element I in G such that for any x in G

$$I * x \;=\; x * I \;=\; x.$$

(5) Every member of G has an inverse: if x is in G then there is an element y of G such that

$$x * y \;=\; y * x \;=\; I.$$

We have already met several examples of groups. If G is the set of all symmetries of a given figure in the plane, and * is the operation of applying two symmetries in succession, the result is a group. Or if G is the set of non-zero rational numbers and * is ordinary multiplication, the result is a group. Or if G is the set of integers and * is ordinary addition, the result is a group. These two examples of number groups both have operations which are *commutative*, i.e.

$$x * y \;=\; y * x$$

for all elements x, y of the group. But this need not necessarily hold for groups in general. For instance, it is not true for the equilateral triangle group. If you look back at the table for this group you will see that $x * v = y$, but $v * x = z$. Groups in which * is commutative are often referred to as *Abelian* (after the Norwegian mathematician Abel, mentioned earlier).

Though there is no need for the operation * of a group to be any familiar operation, say an arithmetical one, if it is then it is usual to call it by its usual name (multiplication, addition, or whatever it may be). But if the operation is unknown or has no common name, it is usual to refer to * as the *multiplication* of the group, and to $a * b$ as the *product* of a and b in the group. This is purely a convenience and on no account should any inference be drawn from such usage.

Because there are so many different examples of groups, the group concept is a very powerful one – and not only in mathematics (where groups
crop up all over the place) but in other subjects as well. The periodicities of
a crystal, the symmetries of atoms, and the interactions of elementary
particles all involve groups. Anything that can be said about groups in
general will be true for any particular group. But just how does the mathematician go about establishing the properties of an *arbitrary, abstract*
group? The answer is that everything has to be done by means of rigorous
mathematical proofs, starting from the definition of a group.

As an example, we shall prove that any element of a group must have
exactly one inverse. (This has to be established before the notation x^{-1} can
be used for *the* inverse of *x*.) Condition 5 in the definition of a group
guarantees that every element of a group has *at least one* inverse, but does
not preclude the existence of more. Of course, for each of the examples of
groups given earlier it is obvious that no element has more than one
inverse, but this does not help us here: what is required is a proof which will
apply to all cases, including possible examples of groups which we may not
have considered before.

Here then is the proof. Let G be any group, and let *x* be any member of
G. Let *y* and *z* be two inverses of *x*. The aim is to prove that $y = z$. Being
inverses of *x*, both *y* and *z* satisfy the requirement of Condition 5:

$$x * y = y * x = I, \tag{3}$$

$$x * z = z * x = I. \tag{4}$$

Applying Condition 4 to *y* gives

$$y = I * y.$$

So, using Equation (4),

$$y = (z * x) * y.$$

Using Condition 3 it follows that

$$y = z * (x * y).$$

So, using Equation (3),

$$y = z * I.$$

Thus, applying Condition 4 to z gives

$$y = z.$$

That completes the proof. Notice how all the structural conditions imposed on a group by the definition are required in order to carry out this proof.

As you might imagine, most proofs in *group theory* are rather more complicated than the above rather easy example (and consequently are not usually written out in quite so much detail), and often they involve other concepts, but they always share the characteristic of consisting entirely of logical steps based only on the initial assumptions.

More Examples of Groups

The symmetry groups considered above were all connected with plane figures, but the same ideas apply to solid shapes in three dimensions. The cube, for example, has 24 rotational symmetries (rotation about an *axis* now, not about a point as in two dimensions). To see this, note that any vertex of the cube may be moved to any other, and that the edges leading to that vertex may be rotated in three ways. If reflections are included (reflections in a *plane* now, not a line) the cube has a total of 48 symmetries.

The *dodecahedron*, which consists of 12 identical regular pentagons fitted together to form a ball-like solid (see Figure 29), has 60 rotational symmetries (120 symmetries when reflections are included). In both the cube and the dodecahedron the rotational symmetries on their own form a group which 'sits inside' the group of *all* symmetries of the figure. Mathematicians would say that the rotational symmetries form a *subgroup* of the entire symmetry group in each case.

The rotational symmetries of the dodecahedron form the smallest non-commutative *simple* group (see later), and it was the simplicity of this group, together with the fact that it has a non-prime number of elements, that Galois used to show that the general quintic polynomial equation could not be solved by radicals.

So far we have seen examples of groups which are infinite (for example addition of the integers) and others which are finite (for example the various symmetry groups). Most of our attention in this chapter will now

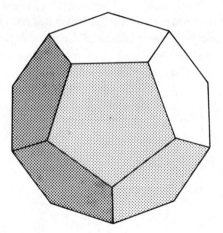

Figure 29. The dodecahedron.

be focused on finite groups. Matrices provide examples of both finite and infinite groups.

A *matrix* is just a rectangular array of numbers (which may be either rational or real as far as our examples are concerned), usually drawn enclosed in brackets, for example:

$$\begin{bmatrix} 1 & 3 & \frac{3}{4} \\ 2 & \frac{1}{4} & -5 \end{bmatrix}.$$

They can be any size, but we shall be concerned exclusively with *square matrices,* where the number of rows is the same as the number of columns (that number being known as the *order* of the matrix). Thus an example of a (square) matrix of order 2 is

$$\begin{bmatrix} 21 & -5 \\ 3 \cdot 8 & 20 \end{bmatrix}.$$

Matrices have their own arithmetic. The rule for adding two matrices (of the same order) is straightforward: you simply add corresponding entries. Thus

$$\begin{bmatrix} 1 & 3 \\ -2 & 6 \end{bmatrix} + \begin{bmatrix} 2 & 5 \\ 3 & 1 \end{bmatrix} = \begin{bmatrix} 3 & 8 \\ 1 & 7 \end{bmatrix}.$$

Multiplication is a little more complicated. Briefly what you do is multiply the rows of the first matrix by the columns of the second, term by term, adding the answers as you go. For matrices of order 2 this is perhaps best explained by means of an algebraic example followed by a numerical one:

$$\begin{bmatrix} a & b \\ c & d \end{bmatrix} \times \begin{bmatrix} v & w \\ x & y \end{bmatrix} = \begin{bmatrix} (av + bx) & (aw + by) \\ (cv + dx) & (cw + dy) \end{bmatrix},$$

$$\begin{bmatrix} 1 & 3 \\ -2 & 5 \end{bmatrix} \times \begin{bmatrix} 2 & 4 \\ 3 & 1 \end{bmatrix} = \begin{bmatrix} (2 + 9) & (4 + 3) \\ (-4 + 15) & (-8 + 5) \end{bmatrix}$$

$$= \begin{bmatrix} 11 & 7 \\ 11 & -3 \end{bmatrix}.$$

A second example will show that matrix multiplication is not commutative (though addition obviously is):

$$\begin{bmatrix} 2 & 4 \\ 3 & 1 \end{bmatrix} \times \begin{bmatrix} 1 & 3 \\ -2 & 5 \end{bmatrix} = \begin{bmatrix} (2 - 8) & (6 + 20) \\ (3 - 2) & (9 + 5) \end{bmatrix}$$

$$= \begin{bmatrix} -6 & 26 \\ 1 & 14 \end{bmatrix}.$$

'Why such a complicated definition of multiplication?' you might ask. 'Why not simply multiply corresponding entries, much as in addition?' Well, the point is that mathematicians developed and studied matrices with certain applications in mind (in particular the solution of large systems of simultaneous linear equations) and those applications required the definitions given. Matrix arithmetic is so important nowadays that any computer system that is aimed at scientific or commercial users is supplied with software to handle it as a matter of course. Indeed, matrix arithmetic is probably the one numerical task most often performed by present-day computers.

The definitions of addition and multiplication of matrices of order 3 or higher are similar to those for order 2 given above, and practically everything that will be said in what follows applies (with only the obvious modifications) to matrices of any size, but for clarity we shall continue to concentrate on matrices of order 2.

The matrices of order 2 (or indeed of any order) form a group as far as addition is concerned. The sum of two matrices (of order 2) is a matrix (of order 2), the addition is associative, there is an identity matrix

$$\begin{bmatrix} 0 & 0 \\ 0 & 0 \end{bmatrix},$$

and the inverse of any matrix is obtained by putting a minus sign in front of all its entries. This group is in fact commutative.

How about multiplication – does that give rise to a group? Certainly the product of two matrices (of order 2) is again a matrix (of order 2), and matrix multiplication is associative. (This is not immediately obvious, but if you work it out algebraically you will see that it is true.) There is an identity element, namely the matrix

$$\begin{bmatrix} 1 & 0 \\ 0 & 1 \end{bmatrix},$$

which leaves any matrix unaltered when multiplied by it. Turning to the question of inverses, by straightforward calculation you can check that

$$\begin{bmatrix} a & b \\ c & d \end{bmatrix} \times \begin{bmatrix} d/H & -b/H \\ -c/H & a/H \end{bmatrix} = \begin{bmatrix} 1 & 0 \\ 0 & 1 \end{bmatrix},$$

where $H = ad - bc$, and similarly with the two matrices on the left-hand side swapped around. So the matrix

$$\begin{bmatrix} a & b \\ c & d \end{bmatrix}$$

will have as its inverse the matrix

$$\begin{bmatrix} d/H & -b/H \\ -c/H & a/H \end{bmatrix},$$

provided this second matrix exists! The only thing that can go wrong is if the quantity H turns out to be zero. (Remember that division by zero is never possible.) Matrices for which this number H is non-zero do have inverses. Such matrices are called *invertible* (or sometimes *non-singular*). Matrices where the number H is zero do not have inverses and are called *singular* (or *non-invertible*).

Because of the existence of singular matrices, which have no inverses, the matrices do not form a group under multiplication. But if you consider only the invertible matrices you do get a group. Is this obvious? Not quite. Associativity is no problem, since it holds for all matrices, whether invertible or not. Since the identity matrix is invertible (and hence a member of the collection being considered), there is an identity element for the group. Inverses? Of course all the members of the chosen collection do have inverses, but you have to check that these inverses *themselves lie in the collection*. This is an easy matter, but has to be done. And then there is one final thing to check. You have to know that when two invertible matrices are multiplied together, the resulting matrix is itself invertible (i.e. still in the collection). Again this is simply a matter of checking the algebra (try it for yourself).

The invertible matrices form a group that is obviously infinite. (It is not commutative, as indicated earlier.) Some very important groups we consider later in this chapter consist of finite subgroups of the invertible matrices.

Another important class of groups is the *clock groups*, the most obvious example of which is provided by the familiar twelve-hour clock. Take as the set the integers from 1 to 12, with the 'clock addition' operation, where 12 counts as a 'zero' and counting past 12 takes you back to the beginning again. So, for example, 5 plus 5 is 10, 7 plus 8 is 3, 11 plus 11 is 10, and 7 plus 12 is 7. This gives a group in which the identity element is 12, and the inverse of any member of the group is the difference between it and 12; so 7 is the inverse of 5, 9 the inverse of 3, and so on.

There is nothing special about the number 12 here. Any number will do. The clock group of order 10 is the structure that lies behind the decimal number system, the clock group of order 24 corresponds to the twenty-four-hour clock, the clock group of order 60 is connected with the measurement of time, and the one of order 360 is related to the measurement of angles.

The mathematician's name for a clock group is a *cyclic group*, so called because the elements of such a group go round in cycles (like a clock). For instance, the cyclic group of order 3 has the table

+	1	2	3
1	2	3	1
2	3	1	2
3	1	2	3

Simple Groups

One of the primary aims in any branch of science is to identify and study the 'basic objects' from which all other objects are constructed. In biology these are the cells (or possibly the molecules), in chemistry the atoms, in physics the fundamental particles (currently the quarks). So too in many branches of mathematics. The classic example is number theory, where (according to the fundamental theorem of arithmetic described in Chapter 1) the prime numbers are the basic building blocks. In each of these examples the basic objects of the theory are *structurally simple*, in the sense that (from the point of view of the theory) they cannot be decomposed into smaller entities of the same kind. (The atoms cannot be broken apart by chemical means, the prime numbers cannot be broken apart by division, and so on.)

The fundamental building blocks in group theory are the *simple groups*. To explain just what these are, and how any given (finite) group can be split into component simple groups, we require the concept of a *telescopic image* of a group. Roughly speaking, when you make a telescopic image of a group G, what you get is a sort of 'scaled-down' version of G. The group operation $*$ of G is reflected in the telescopic image, though in a reduced form. It is a bit like looking at an object through the wrong end of a telescope: the main features of the object are preserved, but it appears smaller and many features may no longer be distinguishable.

To be a little more precise, if you start with a group G, then to form a *telescopic image*, G', of G you have to associate with each element a of G an element a' of G' (called the *image* of a) in such a way that for any pair of elements a, b of G with images a', b', respectively, the product of a' and b' in G' has to be the image of the element $a * b$ of G. (Thus G' preserves the structure of G.) There is nothing here to prevent several elements of G having the same image in G', and indeed it is this 'collapsing' or 'coalescing' of elements that accounts for the reduction in size when you go from G to G'. Mathematicians refer to telescopic images as *homomorphic images*.

Every group G has at least two telescopic images. One of these is G itself, where each element of G is its own image. This situation trivially satisfies the requirements for being a telescopic image, though it is clearly an extreme case. The other telescopic image which any group G possesses lies at the other extreme. It is the *point image* of G, namely the group which has only one element, the identity element *e*. Notice that the definition of a group allows for an identity element to form a group on its own, albeit a rather trivial one, with the multiplication 'table'

$$e * e = e.$$

In the point image, every element of G has the same image, namely *e*, so in this case also the requirements of a telescopic image are satisfied.

Clock groups provide examples of groups which have other telescopic images besides the two trivial ones mentioned above. For instance, let G be the clock group of order 24 and let G' be the clock group of order 12. For each number *n* from 1 to 12 in G, its image *n'* is *n* itself. For *n* between 13 and 24 (inclusive), *n'* is *n* − 12. Then G' is a telescopic image of G. For example, take the elements 7 and 18 in G. Their images in G' are 7 and 6, respectively. The sum of 7 and 6 in G' is 1. According to the definition of a telescopic image, this should be equal to the image of the sum of 7 and 18 in G. Well, in G, 7 and 18 have the sum 1. And the image of 1 in G' is indeed 1. (Notice that the process of going from G to G' is just the usual method of converting time from the twenty-four-hour clock to the twelve-hour clock.)

Observe that in the above example it was important that 12 (the order of G') divides 24 (the order of G). A clock group of prime order has no telescopic images other than itself and the point image. This provides a family of examples of the central notion of this chapter:

> A *simple group* is a group whose only telescopic images are itself and the point image.

By means of telescoping, every finite group may be split into a unique set of simple groups in much the same way that a composite number is decomposed into its prime factors (see Chapter 1). Indeed, the analogy goes further: the number of elements in each of these simple-group components is a factor of the number of elements of the original group, and the product of all these numbers is equal to the number of elements in the original group. There, however, the analogy stops. For one thing, the simple-group components of a group may contain a composite number of elements. (As mentioned earlier, the rotational symmetries of the regular dodecahedron form a simple group of order 60.) Also, whereas the product of all the prime numbers in a given set is a unique number, a given set of simple groups can often be combined in different ways to form quite distinct groups.

The Classification Problem

Having identified the simple groups as being the 'fundamental particles' of finite group theory, mathematicians set about trying to impose some form of classification on the simple groups. Roughly speaking, they wanted to be able to say just which groups are simple and which are not. Of course, the definition of simplicity itself gives some form of answer: the simple groups are those that have only two telescopic images. But that is not the sort of answer that was required. An acceptable solution would provide a description of the actual structures which, considered as groups, turned out to be simple. This could be in the form of a general pattern, giving rise to a family of groups, or a description of an individual 'one-off' group.

For example, one easy result is that all clock groups of prime order are simple (and clock groups of composite order are not). In fact these are the only examples of commutative simple groups, so already we have a complete classification of all the *commutative* simple groups (in the form of a 'regular' family). It was the classification of the non-commutative groups that took all the effort. Over the years from around the 1940s, as mathematicians began to work towards the classification theorem (though not necessarily with that goal in mind), several infinite 'regular' families of simple groups were discovered. In the end a total of 18 such families were found, including the family of all clock groups of prime order mentioned above, and another easily described family – see later. Also found were a number of highly irregular 'one-off' groups that did not fit into any known pattern. The first five of these strange *sporadic* simple groups, as they came to be called, had been found by Émile Mathieu in the 1860s. The smallest of Mathieu's groups has exactly 7920 elements, the largest 244 823 040. It was not until a century later, in 1965, that the sixth sporadic group was discovered by Zvonimir Janko. Janko's group, which has 175 560 elements, consists of a certain collection of matrices of order 7 (with matrix multiplication as the group operation). The way in which this group was discovered is indicative of the kind of work which led to the discovery of an eventual collection of 26 sporadic simple groups. It arose as a result of an examination of the 17th of the regular families of simple groups, a family which was found by Rimhak Ree in 1960, and nowadays known as the Ree family.

As will be explained presently, with every simple group are associated certain smaller groups which provide information about the structure of the simple group. Called *centralizers of involution*, their precise definition will be given later, but for the Ree groups they consist of matrices of order 2 whose entries come from a finite set of numbers of size equal to some odd-numbered power of 3. (If the odd power of 3 is 1, the finite set of numbers turns out to be just 1, 2, and 3.) As part of an attempt to prove an early, restricted form of classification theorem, it was necessary to show that the Ree groups are the only simple groups whose associated centralizers of involution consist of matrices of order 2 whose entries come from a finite set of numbers of size equal to an odd power of some prime number p. Evidence suggested that the prime number p here had to be 3, and this was eventually proved to be true, except in the single case where p is 5 and the odd power is 1. It was this exceptional case that Janko set out to investigate. His intention was to eliminate this one remaining obstacle by proving that there are no simple groups of the type concerned where the number set involved has size 5^1 (i.e. 5). He did not succeed in this goal, but he did manage to obtain a rather curious result. He proved that if there were such a simple group it would have to contain exactly 175 560 elements. Such a precise result suggested that there had to be an actual group hidden somewhere in the background, and after an immense amount of hand calculation Janko succeeded in finding it. Thus was the sixth sporadic group discovered. In honour of Janko it was named J_1.

By using similar techniques with families of simple groups other than the Ree family, Janko soon found evidence for the existence of two further sporadic simple groups, one with 604 800 elements, the other with 50 232 960. He could not, however, actually find such groups. The smaller of the two, J_2, was eventually found by Marshall Hall Jr and David Wales, and the larger one, J_3, was rooted out by Graham Higman and John McKay (using a computer to perform the calculations).

In a more or less similar fashion, the years that followed saw the discovery of several more sporadic groups until, in 1980, the last of the 26 such 'one-off' groups was constructed (its existence having been suspected since 1973) by Robert Griess. It is by far the largest of the sporadic groups, a fact which earned it its name of 'the Monster'. For the record, the number of elements in the Monster is:

$$808\,017\,424\,794\,512\,875\,886\,459\,904\,961\,710\,757\,005$$
$$754\,368\,000\,000\,000.$$

(That is roughly 8 followed by 53 zeros.) It consists of a certain collection of matrices (with entries taken from the complex numbers) of order 196 883. It is worth noting that Griess performed all the necessary calculations to determine the Monster by hand. The fact that the group was 'cooperative' enough to allow such an approach to succeed caused Griess to rename his group 'the Friendly Giant'.

The discovery of the Monster was one of the last steps in the proof of the classification theorem. It is now known that the finite simple groups consist of the groups which make up the 18 regular, infinite families of groups (the first of which is the family of clock groups of prime order), together with the 26 sporadic groups, *and no more*. This is the result which took up 500 articles and almost 15 000 pages in mathematical journals.

The Eighteen Families and the Odd Ones Out

Very often in mathematics progress is made by the formulation of a proposed theorem followed by its proof. With the classification theorem this was not at all the case. Until this theorem had been proved, there was no way of knowing even the size of the problem. There might, for instance, have been many more than 26 sporadic groups, possibly even infinitely many, which would have meant that the goal being sought could never be achieved. The greater part of the work was done more on the basis of 'let's find out about simple groups', rather than as a determined push towards a stated theorem. This makes it difficult to say exactly when the work which led to the final classification began. In his address to the 1954 International Congress of Mathematicians in Amsterdam, Richard Brauer proposed a method for trying to classify the simple groups (of even order, though this restriction ultimately proved to be redundant), and this could count as one starting point. Another, possibly less debatable starting date would be 1972, the year in which Daniel Gorenstein gave a series of lectures at the University of Chicago outlining a 16-step programme that ought to lead to the eventual solution of the classification problem. In some respects the final assault was made possible by a key result obtained by Walter Feit and John Thompson in 1962, and this too could be regarded as 'the beginning of the end'. At any rate, in order to say any more it is necessary first to say something about the nature of the groups that appear in the complete solution.

The first of the 18 regular families has already been mentioned: the family of all clock groups of prime order. The second family may likewise be easily described. For any whole number n greater than 4, the group of all even permutations of n symbols is a simple group, and the collection of all such groups forms the second family. What is an *even permutation of n symbols?* Consider $n = 4$ (the first interesting case). Take four symbols, say the letters A, B, C, D. In alphabetic order these letters form a 'word', ABCD. By repeatedly swapping pairs of letters it is possible to rearrange these four letters into any one of $4 \times 3 \times 2 \times 1 = 24$ different 'words' (or orders). Any such rearrangement is called a *permutation* of ABCD. It is an *even* permutation if it is obtained by an even number of swappings, an *odd* permutation if the number of swaps is odd. For instance, CBDA is an even permutation, being obtained from ABCD by first swapping A and C and then swapping A and D; BACD is an odd permutation, being arrived at from ABCD by swapping just the single pair A and B.

Happy so far? Well now we shall do what we did with symmetries, and think of *permutations* not as the final ordering of the letters, but rather as the sequence of swaps which achieves that final ordering. This means that we can think of *combining* two permutations to form a single permutation: if a and b are permutations (i.e. sequences of swaps) of ABCD, then $a * b$ is the permutation which consists of first performing the a swaps and then the b swaps. For example, if a swaps A and C and then C and D, and if b swaps A and B, then, starting from ABCD, you get

$$a \text{ transforms ABCD to DBAC,}$$

$$b \text{ transforms DBAC to DABC,}$$

$$a * b \text{ transforms ABCD to DABC.}$$

Obviously the operation $*$ is associative. The identity permutation e, which changes nothing, acts as an identity operation, i.e.

$$a * e = e * a = a, \quad \text{for any } a.$$

And the inverse of any permutation obviously consists of the same swappings performed in the opposite order, so if a swaps A, C and then C, D, then a^{-1} swaps C, D and then A, C. (Check for yourself that $a * a^{-1} = a^{-1} * a = e$.)

Thus the permutations of the four letters A, B, C, D constitute a group. The even permutations on their own also form a group, a subgroup of the group of all permutations consisting of exactly half the elements. (This is a simple consequence of the fact that the sum of two even numbers is again an even number, so the ∗ product of two even permutations is another even permutation.) This smaller group is called the *alternating group* of degree 4. It turns out to be an exact copy of the group of all rotational symmetries of a regular tetrahedron (i.e. both groups have the same table, and so to all intents and purposes are the same group).

So much for $n = 4$. The same considerations lead to an alternating group of degree n for any n greater than 2. (If there are only two symbols then there is only one non-trivial permutation, which is odd, so the corresponding alternating group would be the one-point group.) For $n = 3$ the permutation group has $3 \times 2 \times 1 = 6$ elements, and the alternating group is a mirror image of the clock group of order 3. (If a is the even permutation of ABC which swaps A, B and then A, C, then a sends ABC to BCA and $a \ast a$ sends ABC to CAB, and these are the only even permutations there are besides the identity. $(a \ast a) \ast a$ takes ABC to ABC, and the clock has travelled round once.)

For any n greater than 4, the alternating group of degree n is a simple group. This is what lies behind the insolubility (by radicals) of all polynomial equations of degree greater than 4. But wait a moment. Wasn't there an earlier remark to the effect that it was the simplicity of the group of rotational symmetries of the regular dodecahedron that led Galois to his conclusion on quintics? Indeed it was. The dodecahedron group is an exact copy of the alternating group of degree 5.

After the prime clock groups and the alternating groups of degree greater than 4, the remaining 16 regular families are harder to describe in an account such as this. They are all groups of matrices of appropriate sizes. In some cases the families were first described in terms of the matrices involved; in others the family was first defined in other terms and only after considerable effort was a matrix description obtained.

What then of the way in which the classification problem was solved? Many of the regular families were known by the turn of the century, as were Mathieu's five sporadic groups. From the observation that all the known non-commutative simple groups contained an even number of elements, Burnside had conjectured that the same would be true of all non-commutative simple groups, however many there were and whatever other properties they might have. In 1962, Burnside's conjecture was proved correct by Walter Feit and John Thompson of the University of

Chicago, a result for which they were awarded the Cole Prize in Algebra in 1965. Giving a foretaste of the extreme length of the eventual proof of the complete classification theorem, the proof of the Feit–Thompson theorem filled an entire 255-page issue of the *Pacific Journal of Mathematics*. (Mathematical journals such as this typically contain twenty or thirty papers on widely differing topics.)

With the Feit–Thompson theorem available, the way was suddenly opened up for an advance towards the classification theorem along the lines outlined by Brauer in his 1954 lecture mentioned earlier. There are, you see, two aspects to the problem. One is to identify the simple groups (or families thereof) that are required for the classification. Aside from one remaining family and a few sporadic groups, this step had been accomplished by 1960 (though at the time this was not at all clear). The other step is to prove that every simple group does indeed fall into one of the given categories. This is where a great deal of the complexity of the proof arises. The problem is that you have to start with a quite arbitrary simple group (i.e. all you know is that it *is* a simple group), and then somehow show that it is (an exact copy of) a member of one of the regular families, or else one of the listed sporadic groups. What Brauer suggested was a way of attacking this problem.

His idea was to concentrate on the elements a of the group (other than the identity element e) for which $a * a = e$. Such group elements are called *involutions*, and it is easy to show that any group with an even number of elements must contain at least one involution. (Try this yourself. All you need to know about groups is the definition given earlier. The solution is so neat and concise that its discovery is well worth the effort it might take you to find it.) By virtue of the Feit–Thompson theorem, it follows that every non-commutative simple group will contain involutions.

What Brauer did (and remember, this was before the Feit–Thompson theorem, but after Burnside had made the conjecture that became their theorem) was to calculate centralizers of involutions in several of the known regular families. Centralizers? The *centralizer* of an element g of a group G is the set of all elements a of the group for which $a * g = g * a$. If G is commutative, then the centralizer of any element is just G itself, of course, but in other cases this need not be so. What *is* true – and is quite straightforward to verify – is that the centralizer of any element of G is a subgroup of G. Brauer's observations were encouraging. The centralizers-of-involutions groups all had the same general structure as the original simple group, though in an embryonic form. This led him to suspect that it might be possible to reconstruct the entire group from a knowledge of these centralizer groups, and in certain special cases he was able to confirm this suspicion.

Not only does Brauer's work lie behind the discovery of many of the sporadic groups (the three Janko groups have been mentioned already), but it provides the beginnings of a way of pinning down an arbitrary given simple group within the proposed classification. First, show that the centralizer of an involution in the given group closely resembles the centralizer of an involution in one of the known simple groups in the classification scheme. Then try to extend this highly localized 'tie-up' to a complete equivalence. This last step is by no means easy: the centralizer of an involution is just a tiny part of the entire group, so it is a bit like trying to deduce the theme of an entire jigsaw puzzle from just one piece.

Following Brauer's approach was a part of the 16-step programme outlined by Gorenstein in his 1972 Chicago lectures. Gorenstein himself thought that the task could be accomplished by the end of this century. Most of his audience thought this wildly optimistic. They all reckoned without the presence in the audience of a young mathematician who had just completed his graduate studies. Starting with a key result known as the *component theorem*, Michael Aschbacher of the California Institute of Technology stormed his way through the problem, proving one startling result after another, with the result that by 1980, just eight years after Gorenstein's lectures, Solomon was able to supply that one small step for mathematics that marked the completion of the proof. (Aschbacher won the 1980 Cole Prize for Algebra for his work.)

Along the way the various remaining sporadic groups were uncovered. As with all the regular families except the first two, these all consist of certain sets of matrices, and in some cases computers were required to perform the necessary computations. Given that these 26 groups are, by their very scarcity amongst the infinitude of all simple groups, obviously quite special, it should come as no surprise to learn that they have connections with other branches of mathematics. For instance, the discovery in 1968 by John Conway of Cambridge University of the three sporadic groups which now bear his name was based on Leech's lattice, a mathematical structure that arose in work on the design of error-correcting codes (methods for encoding transmitted information so that distortions and occasional losses can be compensated for). And two of Mathieu's sporadic groups are related to the Golay error-correcting code that is often used for military purposes. Connections such as these provide some cause for interest in the classification theorem, but its main 'claim to fame' outside group theory itself undoubtedly lies in the incredible length of the proof. The last word on the subject I shall leave to the man who played such a large part in obtaining the final proof. Michael Aschbacher, looking back on the proof soon after its completion in 1980, wrote:

> 'Much of the mathematics involved has been produced only recently and will no doubt be improved once there is time for the techniques to sink in. Still, it is hard to imagine a short proof of the theorem. I personally am sceptical that a short proof of any kind will ever appear.
>
> Long proofs disturb many mathematicians. For one thing, as the length of a proof increases, so does the possibility of an error. The probability of an error in the proof of the classification theorem is virtually 1. On the other hand the probability that any single error cannot be easily corrected is virtually zero, and as the proof is finite, the probability that the theorem is incorrect is close to zero. As time passes and we have an opportunity to assimilate the proof, that confidence level can only increase.
>
> It is also perhaps time to consider the possibility that there are some natural, fundamental theorems which can be stated concisely, but do not admit a short, simple proof. I suspect that the classification theorem is such a result. As our mathematics becomes more sophisticated, we may encounter such theorems more frequently.'

Suggested Further Reading

A gentle introduction to group theory can be found in Chapter 7 of *Concepts of Modern Mathematics*, by Ian Stewart (Pelican, 1981).

At a higher level, a detailed account of the subject-matter of this chapter is given in *Finite Simple Groups*, by Daniel Gorenstein (Plenum, 1982).

The definitive account of the classification of the finite simple groups is contained in *The Classification of Finite Simple Groups*, Volumes 1 and 2, by Daniel Gorenstein (Plenum, 1982).

6   Hilbert's Tenth Problem

A Historic Gathering

In August 1900 the world's best mathematicians gathered in Paris for the Second International Congress of Mathematicians (an event which, except for periods of war, has continued to be held every four years at different venues around the globe). Amongst them was the 38-year-old professor from the University of Göttingen, David Hilbert. As one of the leading mathematicians of the time, Hilbert was due to deliver one of the keynote addresses to the meeting. The day fixed for his lecture was 8 August.

As the meeting was being held in the very first year of the twentieth century (indeed, it had been brought forward a year in order to achieve this), Hilbert chose to use his lecture not to look back over some recent work (which is the usual format for such talks), but rather to point the way towards the future.

'We hear within us the perpetual call: there is the problem,' he cried. 'Seek its solution. You can find it by reason, for in mathematics there is no *ignorabimus* [we shall not know].' To emphasize this call he presented to the meeting not one but a list of twenty-three major unsolved problems – problems whose solutions, if found, would each mark a significant advance in mathematical knowledge. Most of these problems were (or became) known by specific names, such as the *continuum problem* (the first on Hilbert's list – see Chapter 2) or the *Riemann problem* (see Chapter 9), but one in particular became universally known by its position on Hilbert's list: the tenth.

Hilbert's tenth problem has its origins in an algebra textbook, the *Arithmetica*, written around AD 250 by Diophantus of Alexandria (see Chapter 8). In accordance with the kinds of problem considered in this tract, present-day mathematicians use the name *Diophantine equation* to refer to any equation in one or more variables, with integer coefficients, where a solution is sought *consisting entirely of integers*. It is this last condition which makes the mathematics of Diophantine equations radically different from the solution of equations in the real numbers (or possibly the complex numbers). (Actually the nomenclature is a bit confusing at first. The adjectival use of the word 'Diophantine' refers not to the equation so much as to the kind of solution which is sought. Thus the equation

$$3x^2 - 5y^2 + 2xy = 0$$

is referred to simply as an 'equation' if real-number solutions are sought, but as a 'Diophantine equation' if only integer solutions are required.)

Solving Diophantine equations is quite different from solving the same equations in the real numbers. For instance, if you take the equation

$$x^2 + y^2 = 2 \tag{5}$$

and regard it as a regular equation for real numbers, then there are infinitely many solutions. Given any real number r between $-\sqrt{2}$ and $+\sqrt{2}$, if you take

$$s = +\sqrt{2 - r^2}$$

then $x = r$, $y = s$ gives you a solution. Considered as a Diophantine equation, however, there are just four solutions:

$$x = +1, \quad y = +1; \quad x = +1, \quad y = -1;$$
$$x = -1, \quad y = +1; \quad x = -1, \quad y = -1.$$

If you change the equation just slightly, say to

$$x^2 + y^2 = 3, \tag{6}$$

then there are still infinitely many real solutions but no integer solutions at all. As a Diophantine equation, Equation (6) cannot be solved. So what is the difference between Equations (5) and (6)? More generally, is there a way of telling whether or not any given Diophantine equation can be solved? For instance, would it be possible to write a computer program which, given a Diophantine equation, tells you whether there is a solution or not? This is essentially the question Hilbert asked as the tenth in his list of problems. Its solution in 1970 by the 22-year-old Russian mathematician Yuri Matyasevich came only after a great deal of work had been done on the problem, stretching back to the 1930s and encompassing results in mathematical logic, computation theory, and algebra.

Diophantine Equations and the Euclidean Algorithm

The simplest kind of Diophantine equation is a linear one, with one unknown. In fact the only information we have of Diophantus' life is itself in the form of such an equation. A fourth-century problem states that his boyhood lasted for one-sixth of his life, his beard grew after a further one-twelfth, and after one-seventh more he married; his son was born 5 years later and lived to half his father's age, dying 4 years before him. If you let x denote the age at which Diophantus died, this information gives us the equation

$$\frac{1}{6}x + \frac{1}{12}x + \frac{1}{7}x + 5 + \frac{1}{2}x + 4 = x,$$

which has the solution $x = 84$. (Strictly speaking this is not a Diophantine equation since the coefficients are not integers, but by multiplying through by the least common multiple of all the denominators in the coefficients, an equation with integer coefficients is obtained.) Whether or not Diophantus really did live to 84, the fact remains that the solution of a linear Diophantine equation in one unknown is a trivial matter. The equation

$$ax = b$$

has an integer solution if, and only if, a divides b (exactly), in which case the solution is the integer b/a. This condition is so simple that it is easy to write a computer program which will tell you at once whether such a Diophantine equation has a solution.

What about linear Diophantine equations in two unknowns? Here again there is a simple way of finding out whether there is a solution or not. To see if the equation

$$ax + by = c$$

has an integer solution, you first calculate the highest common factor, d say, of a and b. If d divides c, then there is a solution; if d does not divide c there is no solution.

For example, does the equation

$$6x + 15y = 12$$

have a solution? Well, the highest common factor of 6 and 15 is 3, and 3 does divide 12, so there is a solution. (For example $x = 7, y = -2$ solves the equation.)

Notice that, for a given Diophantine equation, determining whether or not a solution exists is not at all the same as *finding* a solution. It may be possible to determine the existence of a solution quite easily, and yet be extremely difficult actually to find one. (Though, conversely, if you know how to find a solution then you know at once that there is a solution! If you can find something, then that thing must exist, whereas things can exist without being found.) For linear Diophantine equations in two unknowns, not only is there a simple way of determining whether a solution exists, there is also a mechanical procedure for finding a solution if there is one. Complete details can be found in most elementary textbooks on number theory.* The key to the solution is the *Euclidean algorithm* for the determination of highest common factors, described below.

Given two numbers x and y, let $x \bmod y$ denote the remainder that is left when x is divided by y (as in Chapter 1). To calculate the highest common factor of two given numbers a and b, with a greater than b, proceed as follows. Let $a \bmod b = r_1$, then let $b \bmod r_1 = r_2$, then let $r_1 \bmod r_2 = r_3$. Continue in this way until a remainder of zero is obtained: $r_{n-1} \bmod r_n = 0$. Then r_n is the highest common factor of a and b.

For example, to find the highest common factor of 133 and 56, you would proceed as follows:

*For example *Elementary Number Theory*, by David Burton (Allyn and Bacon, 1980).

$$133 \bmod 56 = 21,$$
$$56 \bmod 21 = 14,$$
$$21 \bmod 14 = 7,$$
$$14 \bmod 7 = 0.$$

So the highest common factor of 133 and 56 is 7, the last non-zero remainder. (At this stage you might like to verify for yourself that the numbers 81 and 25 have highest common factor 1.)

The procedure just described appeared in Book VII of Euclid's *Elements*, written around $350-300$ BC, which explains why it is nowadays known as the Euclidean algorithm. But what exactly is an 'algorithm'? This question is crucial as far as Hilbert's tenth problem is concerned. Before attempting to answer it, let us look briefly at what else was known about the solution of Diophantine equations when Hilbert gave his lecture.

In fact there was (and still is) very little known. Linear equations with more than two variables can be dealt with by an extension of the Euclidean algorithm method for the two-variable approach, mentioned above. For second-degree equations with one or two unknowns, such as

$$x^2 - 3x + 4 = 0$$

or

$$3x^2 - 5xy + y^2 = 7,$$

an impressive theory worked out by Gauss provides a procedure for determining whether a given equation has a solution or not. (This is the famous quadratic reciprocity theory, mentioned on p. 69.) But except for occasional special cases where clever tricks can be performed, that is more or less all that was known. (A particularly important 'special case' concerns the Diophantine equations

$$x^n + y^n = z^n,$$

where n is at least 2. The existence of solutions to these equations for n greater than 2 is the famous problem of Fermat's last theorem, described in detail in Chapter 8.)

And now, what about that notion of an 'algorithm'?

Algorithms and Turing Machines

Some time around AD 825, a Persian mathematician called al-Khowarizmi wrote a book outlining the rules for performing basic arithmetic using numbers expressed in the Hindu decimal form that we use today (with columns for units, tens, hundreds, and so on, and decimal points to denote fractional quantities). From his name comes the modern word 'algorithm'.

An *algorithm* is a step-by-step method for performing some kind of calculation. Exactly how the instructions are written down or specified is not important. What does matter is that the instructions should be complete and unambiguous, with no room for choice or chance, and should work for all possible starting data, not just particular values. The Euclidean algorithm described in the last section is a good example. The instructions tell you exactly what to do at each step, and the procedure works for any numbers *a* and *b* (with *a* greater than *b* – to allow for *all* cases simply add an initial instruction to write the two numbers with the greater preceding the smaller). Other examples of algorithms are the rules for addition, subtraction, multiplication, and division of numbers written in decimal format that al-Khowarizmi laid down in his book.

Hilbert's tenth problem asks if there is an algorithm to determine whether a given Diophantine equation has a solution. For some particularly simple Diophantine equations there is – there are algorithms for linear and quadratic equations in at most two unknowns, as mentioned earlier. But is there an algorithm which works in all cases? If the answer is yes, then in order to prove this it would be enough to write down the appropriate algorithm. But supposing the answer is no. How do you set about proving that? The notion of a 'step-by-step instruction set', whilst adequate for recognizing a specific algorithm as such, is altogether too vague to enable us to prove that there is no algorithm for performing a certain task. For that a rigorous, mathematical definition is required.

Given present-day computer technology, one could formulate a definition of 'algorithm' in terms of a computer program written in a specific computer language for a specific machine, and this would certainly be precise enough. But some obvious problems arise. For one thing, which language and which machine? And what about limitations on the size of numbers that can be handled by the machine or the amount of memory that is available? It turns out that, provided you are prepared to idealize the situation by removing all limitations on data size, the choice of the language and machine are unimportant as far as the resulting definition of 'algorithm' is concerned. They all give rise to precisely the same collection of calculable functions: a calculation will be possible on one machine with one language if, and only if, it is possible on any other machine using any other language. This is intuitively clear when you realize that, at their most basic level of operation, all that computers do is manipulate 0s and 1s.

In fact there is no need to resort to computer technology in order to obtain a workable definition of 'algorithm'. Several definitions were formulated by mathematical logicians (notable among them Emil Post, Alonzo Church, Stephen Kleene, Kurt Gödel, and Alan Turing) during the 1930s, long before the advent of the computer age. Some quite different approaches were adopted: an 'equational calculus', a calculus of 'recursive functions', and various abstract 'machines'. But in each case the resulting notion of a 'performable calculation' turns out to be the same, so as far as defining the notion of an 'algorithm' is concerned you can choose any of these approaches. You may as well choose the simplest method, which was worked out by the English logician Alan Mathison Turing.

Turing postulated the existence of an abstract computing 'machine', known nowadays as a *Turing machine*. This consists of a read/write *head* through which passes a two-way infinite *tape*, divided into squares (see Figure 30). The squares on the tape can either be blank or carry one symbol from a fixed alphabet (0 and 1 are sufficient, but the exact choice of alphabet is unimportant in the long run). At any one time the head will be in one of a finite number of different *states*. (Two states are sufficient, but the exact number is not an important factor.) The operation of the machine proceeds in a step-by-step fashion, and each step consists of three distinct actions. At any instant the head will be scanning one tape square. What it does depends on the contents of that square and the state of the machine. Depending on these two factors, the machine erases the existing tape character and either leaves the square blank or else writes another (or possibly the same) symbol there, then it moves the tape through the head by one square in either direction, and finally it goes into another (possibly the same) state. The overall behaviour of the machine is determined by an *instruction set* which says – for each state and each possible symbol read – exactly which three actions should be performed. The initial data (if there is any) is supplied by writing it on the tape (according to some coding system, the choice of which is not important), and the machine is set

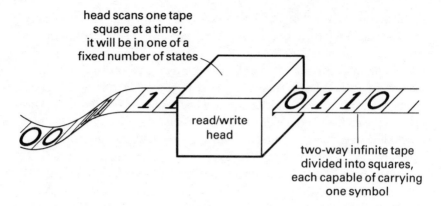

head scans one tape
square at a time;
it will be in one of a
fixed number of states

read/write
head

two-way infinite tape
divided into squares,
each capable of carrying
one symbol

Figure 30. A Turing machine. This hypothetical computing
device was invented by the English mathematician Alan Turing in
the 1930s to provide an abstract framework for the study of computation. Despite its simplicity, it can be shown that any computation, no matter how complicated, can be carried out using a Turing
machine. The Turing machine concept allows a precise definition of
an 'algorithm' as a Turing machine program.

The read/write head is at any instant in one of a fixed, finite
number of states. The action of the machine proceeds in discrete
steps. The actual step at any stage depends upon the current state
of the head and the contents of the tape square being scanned by the
head. The operation of the machine is controlled by a program,
which consists of a table indicating what action follows each combination of head state and tape input character (see Box B).

in motion with the head scanning the first data square. If and when the
computation comes to an end, the machine enters a special *halt* state and
ceases to operate. Any answer that has been produced will then be found
on the tape, starting from the square being scanned when the machine
stops.

In terms of a Turing machine, an *algorithm* consists of a sequence of
instructions that determines the behaviour of the machine in the manner
indicated above. Obviously with such a very simple machine even the most
basic calculation will require a painstakingly detailed 'algorithm' (see Box
B). But the whole point of the concept is that it provides a precise definition
of an algorithm (and of a *computation*) which is simple enough to be handled
mathematically and yet which is adequate for performing any 'algorithmic
calculation'. There is no suggestion that such a device should actually be
built – though numerous enthusiasts have done just that!

Box B: A simple Turing machine program

For this example, the alphabet of tape symbols consists of 0 and 1 only. Positive integers are represented by a consecutive sequence of 1s, with n 1s denoting the number n (so the positive integers are denoted by: 1, 11, 111, 1111, ...). There are five states, labelled 1, 2, 3, 4, and H (the special halt state). The object of the program is to decide if a given integer on the input tape is even or odd. If it is even, the machine should output a 1 and stop; if it is odd, it should output a 0 and stop. This output is to appear on the tape after the integer, with one blank square in between. It is assumed that the input integer is aligned with the head initially scanning its first digit and the rest of the integer to the right.

In the program table below, b denotes the blank tape 'symbol', R means scan the next tape square to the right. More complicated programs would probably involve some movements to the left as well.

Condition		Action		
State	**Input**	**Output**	**State**	**Motion**
1	1	1	2	R
1	b	b	3	R
2	1	1	1	R
2	b	b	4	R
3	–	1	H	–
4	–	0	H	–

continued overleaf

The action of this program for input 3 (i.e. 111 on the tape) is followed below, step by step. The arrow indicates the tape square being scanned, and the current head state is given.

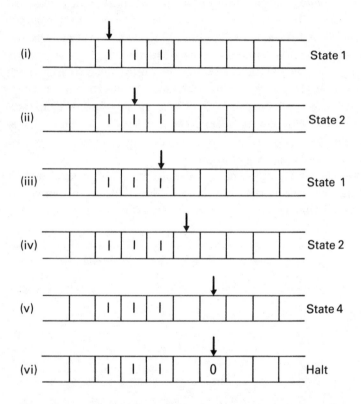

You might like to follow the action of this program for one or two other inputs, or work out programs for performing other simple tasks such as the addition of two positive integers represented in the above manner.

Computable Sets

A crucial notion for the solution of Hilbert's tenth problem is that of a *computable set* of integers. This is a set S of integers for which there is a mechanical (or algorithmic) method of determining which numbers are in S and which are not. In terms of a Turing machine, a set S of integers is said to be *computable* if there is a Turing machine program which, given any integer as input, halts with an output of 1 if the integer is a member of S and halts with an output of 0 if the integer is not a member of S. For example, the set S of even integers is computable. The program outlined in Box B performs the necessary calculation. (Actually, this program deals only with positive integers. To allow for negative integers you need some convention for coding positive and negative numbers, such as using the first symbol as a sign. As an exercise you might like to try to modify the program given in Box B to handle this more general case.)

Notice that in the above definition of a computable set the Turing machine program produces an answer for every input – it cannot, as so often occurs with real programs running on real computers, go into an infinitely recurring loop or start a never-ending search for some data item that does not exist. A weaker notion, that allows for such 'unending calculations', is that of a *listable set* of integers (known to mathematicians as a *recursively enumerable set*). In terms of a Turing machine, a set S of integers is said to be *listable* if there is a Turing machine program which, given any integer as input, halts with an output of 1 if, and only if, the integer belongs to S. If the input integer does not belong to S, the program may halt with output 0, or it may not halt at all. Thus if you run the program on a given input integer N, if it happens that N is a member of S then eventually the program will tell you so, but if N is not a member of S, you might never find this out – the calculation might go on for ever, though you could never be sure that it was not just about to stop. So it is a very one-sided situation.

As you might imagine, there is a close relationship between the two notions of a computable set and a listable set. A set S of integers is computable if, and only if, both S and $\overline{S}$ are listable, where $\overline{S}$ is the *complement* of S (that is, the set of all integers that are not members of S). It is fairly easy to see why this is so. If S is computable, then any program P that testifies to this fact will also testify the listability of S, and to obtain a program which shows that $\bar{S}$ is listable, take the program P and add a final step which replaces an output of 0 by 1, and an output of 1 by 0. Conversely, if both S and $\bar{S}$ are listable, then to obtain a program showing that S is computable, proceed as follows. Let P and Q be programs which give the listability of S and $\bar{S}$, respectively. If you now take two Turing machines, one running program P and the other Q, and if you feed a given integer N simultaneously into both machines, then if N is a member of S the program P will eventually halt with output 1, and if N is a member of $\bar{S}$ the program Q will eventually halt with output 1. So taken together, the two Turing machines give you a mechanical way of deciding whether a given integer is in S or not. Intuitively this tells you that S is computable. To make it precise in terms of the definitions, you have to construct just one Turing machine program that does the job of the two programs P and Q. An obvious way is to write a program R which, for a given integer input, alternately runs P and Q on that input (say 100 steps of each, in turn) until one of these halts with output 1. If it is the P part that does this, then R outputs 1 and halts; if it is the Q part then R outputs 0 and halts. Obviously, program R will testify to the computability of S.

The above result is the best that can be achieved in establishing the relationship between computability and listability. It is definitely not the case that the two notions are the same. Every computable set is, of course, listable, but there are listable sets that are not computable. To construct an example of such a set you need Turing's concept of a 'universal' Turing machine – a single Turing machine program that can simulate the operation of *every* Turing machine program. Turing showed that such a program can be constructed, and though the exact details of the construction are rather technical the underlying idea is a simple one. Start with a listing of all Turing machine programs: $P_1, P_2, P_3, \ldots$. The universal program runs like this. On the input tape you first put a natural number N. When the universal program reads N, it then runs program P_N (any necessary data being found on the tape following N).

Next, to construct a set S that is listable but not computable, let S be the set of all those natural numbers N such that program P_N halts with output 1 when it runs on input N (where $P_1, P_2, P_3, \ldots$ is the listing of all Turing machine programs used to obtain the universal Turing machine program above). Using the universal Turing machine program, it is easy to show that S is listable: simply add to the beginning of the universal program a small program that takes any input integer on the tape and replaces it by two identical copies of itself, leaving the head scanning the start of the first copy. (It is a good exercise to check for yourself how this works.)

To show that S is not computable suppose that, on the contrary, it were computable. Then, as observed above, $\overline{S}$ would be listable, so there will be some program which testifies this fact. This program will have to appear in the list $P_1, P_2, P_3, \ldots$ of all programs, say as P_k. Now, either k will be a member of S or it will not. If k is a member of S, then k is not a member of $\overline{S}$, so as P_k 'lists' $\overline{S}$, P_k cannot halt with output 1 from k. So k does not satisfy the condition which defines S, and cannot be a member of S. Thus if k is in S, then it is not! On the other hand, what happens if k is not in S? It will be a member of $\overline{S}$, so P_k will halt with output 1 on k, so k satisfies the defining requirement of S, which means that k is a member of S. That is, if k is not a member of S, then it is! (Does this seem familiar to you? See Chapter 2, and in particular Russell's paradox and the proof of Cantor's theorem.) Having thus arrived at a contradiction, the inescapable conclusion is that S *cannot* in fact be computable, as was first supposed.

With this last result available to us, it is now possible to describe the solution to Hilbert's tenth problem.

Hilbert's Tenth Problem

David Hilbert did not, in fact, ask directly *if* there was an algorithm to determine whether a given Diophantine equation has a solution. Rather, he asked for such an algorithm to be produced. To quote his own words:

> 'Given a Diophantine equation with any number of unknown quantities and with rational integral numerical coefficients: to devise a process according to which it can be determined by a finite number of operations whether the equation is solvable in rational integers.'

On the other hand, elsewhere in his lecture he says (of problems in general):

> 'Occasionally it happens that we seek the solution under insufficient hypotheses or in an incorrect sense, and for this reason do not succeed. The problem then arises: to show the impossibility of the solution under the given hypotheses, or in the sense contemplated.'

In the case of the tenth problem, this is exactly what happened. What Matyasevich proved in 1970 was that no such algorithm exists.

The first serious attempt to obtain such a result was made by Martin Davis in 1950. His strategy was this (see the example that follows if you find this confusing): prove that for every listable set S there is a corresponding polynomial $P_S(x, y_1, y_2, \ldots, y_n)$, with integer coefficients, such that a positive integer k belongs to S if, and only if, the Diophantine equation

$$P_S(k, y_1, y_2, \ldots, y_n) = 0$$

has a solution. (The degree of P_S is not important, nor is the number of variables involved. When the problem was finally solved along the lines Davis had suggested, it turned out that P_S need not have degree greater than 4 and n need not be greater than 14.)

For example, let S be the set of all positive integers which are not of the form $4k + 2$, for some k. Thus

$$S = \{1, 3, 4, 5, 7, 8, 9, 11, \ldots\}.$$

In fact, S is exactly the set of numbers that can be expressed as the difference of two squares (i.e. numbers of the form $a^2 - b^2$ for numbers a and b). Thus,

$$1 = 1^2 - 0^2, 3 = 2^2 - 1^2, 4 = 2^2 - 0^2,$$
$$5 = 3^2 - 2^2,$$

but there are no numbers a and b for which

$$6 = a^2 - b^2.$$

The general proof goes like this. If n is in S, it must have one of the forms $4k$, $4k + 1$, or $4k + 3$. In the first case

$$n = \left(\frac{n}{4} + 1\right)^2 - \left(\frac{n}{4} - 1\right)^2,$$

and in the other two cases

$$n = \left(\frac{n+1}{2}\right)^2 - \left(\frac{n-1}{2}\right)^2.$$

On the other hand, every square is either a multiple of 4 or else one more than a multiple of 4, depending on whether it is the square of an even number or an odd number, so the difference of two squares can never be two more than a multiple of 4, and hence numbers not in S are not a difference of two squares.

Suppose now that with the set S (which is obviously listable) we associate the polynomial

$$P_S(x, y_1, y_2) = y_1^2 - y_2^2 - x.$$

Then, as you may verify for yourself, a positive integer k will be a member of S if, and only if, the Diophantine equation

$$P_S(k, y_1, y_2) = 0$$

has a solution; that is, if, and only if, there is an integer solution to the equation

$$y_1^2 - y_2^2 - k = 0.$$

Of course, the example above works because of the special property of S mentioned. What Davis set out to do was associate an appropriate polynomial P_S with *every* listable set S. To see why this would imply the non-existence of an algorithm of the kind asked for by Hilbert, suppose, on the contrary, that there were such an algorithm. Let S be a listable but non-computable set of integers, as constructed in the previous section. By our supposition that an algorithm exists to determine the solubility of Diophantine equations, there is a Turing machine program H which halts with output 1 on input k if the Diophantine equation

$$P_S(k, y_1, y_2, \dots y_n) = 0$$

has a solution, and halts with output 0 on input k if it does not have a solution. But because of the relationship between S and P_S, H now testifies that S is computable, contrary to the choice of S. Hence no such program H can exist. In other words, there cannot be an algorithm of the kind requested by Hilbert.

Unfortunately, though the strategy works in principle, Davis was unable to prove that such a polynomial P_S always exists. The key to the problem turned out to lie in some work begun by Julia Robinson. She investigated the kinds of sets that can be defined by Diophantine equations, and she was able to develop various techniques for handling equations whose solutions increased in an exponential fashion. In 1960 she collaborated with Davis and Hilary Putnam to show that if *just one* Diophantine equation could be found whose solutions behaved exponentially in an appropriate sense, then it would be possible to describe every listable set by a Diophantine equation in the manner sought by Davis, and hence to solve Hilbert's tenth problem. But they in turn were brought to a halt by being unable to find such an equation. And it was not until ten years later that Yuri Matyasevich succeeded where the three Americans had failed. He did so by making use of a famous sequence of numbers connected with a twelfth-century problem concerning rabbits.

Fibonacci's Rabbits and Matyasevich's Solution

In 1202 the Italian mathematician Leonardo of Pisa published (under the name Fibonacci, from the Latin 'filius Bonacci' meaning 'son of Bonacci') his book *Liber Abaci*, an influential work which introduced the Hindu–Arabic decimal number system to Western Europe. Amongst the problems considered in the book is the following:

> A man puts a pair of rabbits in a certain place surrounded by a wall. How many pairs of rabbits can be produced from that pair in a year, if the nature of these rabbits is such that every month each pair bears a new pair which from the second month on becomes productive?

If it is assumed that one month elapses before the initial pair starts to produce, that there are no deaths in the rabbit colony, and that each pair continues to produce regularly, it does not take long to see that the number of adult rabbit pairs present month by month is given by the numbers in the sequence

$$1, 1, 2, 3, 5, 8, 13, 21, 34, \dots,$$

which is generated by the simple rule that every number after the first two 1s is obtained by adding together the previous two numbers. Thus, $2 = 1 + 1, 3 = 1 + 2, 5 = 2 + 3, 8 = 3 + 5$, and so on.

This simple sequence turns out to have several interesting properties in its own right, as well as some surprising applications. (For instance, it arises in the theory of computer databases, and also in the study of the computational efficiency of the Euclidean algorithm.) As far as Hilbert's tenth problem was concerned, the importance of the Fibonacci sequence lay in the fact that it exhibits exponential growth. The nth number in the sequence is approximately equal to

$$\frac{1}{\sqrt{5}}\left[\frac{1}{2}(1 + \sqrt{5})\right]^{n}.$$

(The approximation improves as n increases.) This meant that, by virtue of the Davis–Robinson–Putnam result mentioned earlier, in order to solve Hilbert's tenth problem it was sufficient to find a Diophantine equation whose solutions were appropriately related to the Fibonacci numbers. This is exactly what Matyasevich did. To obtain the Diophantine equation he discovered, you first start with the following ten polynomial equations:

$$u + w - v - 2 = 0,$$
$$l - 2v - 2a - 1 = 0,$$
$$l^2 - lz - z^2 - 1 = 0,$$
$$g - bl^2 = 0,$$
$$g^2 - gh - h^2 - 1 = 0,$$
$$m - c(2h + g) - 3 = 0,$$
$$m - fl - 2 = 0,$$
$$x^2 - mxy + y^2 - 1 = 0,$$
$$(d - 1)l + u - x - 1 = 0,$$
$$x - v - (2h + g)(l - 1) = 0.$$

In these equations the values of u and v are related in such a way that v is the $2u$th Fibonacci number, and this is enough to satisfy the requirements of the Davis–Robinson–Putnam result. If you now simply square each of these ten equations and add them all together to obtain one big equation, you get the desired single equation which resolves Hilbert's tenth problem. You also get a great deal more.

Looked at in terms of Hilbert's question, the Davis–Robinson–Putnam–Matyasevich solution is a negative one: it demonstrates that there is no suitable algorithm. But in reality it is a very positive mathematical result. It says that every listable set of integers can be described using a Diophantine equation: if S is a listable set, then there will be a polynomial $P(x, y_1, y_2, \dots, y_n)$, with integer coefficients, such that a number k belongs to S if, and only if, the Diophantine equation

$$P(k, y_1, y_2, \dots, y_n) = 0$$

has a solution.

For instance, the set of primes is a listable set. (In fact it is a computable set. It is a routine matter to write a computer program which tests primality – though, as indicated in Chapter 1, it is not so easy to write one which does this in an efficient way.) Consequently the set of primes may be described by means of a Diophantine equation. By means of a little algebraic manipulation it follows that there is a polynomial $P(x_1, \dots, x_n)$ whose positive values (as the variables $x_1, \dots, x_n$ range over all integers) are precisely the primes. This resolves a long-standing question as to whether the primes could be obtained as the values of a polynomial function. (Though note that not all values of the function are primes – the function also produces negative values, which may or may not be minus primes. But the positive values range over all the primes, and no other positive values occur.)

Unfortunately, the Matyasevich result merely implies the existence of such a prime-generating polynomial. It does not indicate how to construct one, and it took a considerable effort before James Jones, Daihachiro Sato, Hideo Wada, and Douglas Wiens finally found one in 1977. Their polynomial has 26 variables and is of degree 25. It is

$$(k + 2) \{1 - [wz + h + j - q]^2$$
$$- [(gk + 2g + k + 1)(h + j) + h - z]^2$$
$$- [2n + p + q + z - e]^2$$
$$- [16(k + 1)^3(k + 2)(n + 1)^2 + 1 - f^2]^2$$
$$- [e^3(e + 2)(a + 1)^2 + 1 - o^2]^2$$
$$- [(a^2 - 1)y^2 + 1 - x^2]^2 - [16r^2y^4(a^2 - 1) + 1 - u^2]^2$$
$$- [((a + u^2(u^2 - a))^2 - 1)(n + 4dy)^2 + 1 - (x + cu)^2]^2$$

$$- [n + 1 + v - y]^2 - [(a^2 - 1)l^2 + 1 - m^2]^2$$
$$- [ai + k + 1 - l - i]^2$$
$$- [p + l(a - n - 1) + b(2an + 2a - n^2 - 2n - 2) - m]^2$$
$$- [q + y(a - p - 1) + s(2ap + 2a - p^2 - 2p - 2) - x]^2$$
$$- [z + pl(a - p) + t(2ap - p^2 - 1) - pm]^2 \}.$$

(Note the apparent paradox that the formula appears to split into two factors. What happens is that the formula produces only positive values when the factor $k + 2$ is prime and the second factor is equal to 1.)

A suitably positive result with which to end this chapter!

Suggested Further Reading

Both Hilbert's famous 1900 paper to the International Congress of Mathematicians and the solution to Hilbert's tenth problem can be found in the book *Mathematical Developments Arising from Hilbert Problems*, edited by Felix Browder (American Mathematical Society, 1974), Volume 28 in the series 'Proceedings of Symposia in Pure Mathematics'.

7 The Four-Colour Problem

Computer Mathematics Comes of Age

In 1976, two mathematicians at the University of Illinois, Kenneth Appel and Wolfgang Haken, announced that they had solved a century-old problem to do with the colouring of maps. They had, they said, proved the *four-colour conjecture*. This in itself was a newsworthy event. The four-colour problem was, after Fermat's last theorem (see Chapter 8), probably the second most famous unsolved problem in mathematics. But for mathematicians the really dramatic aspect of the whole affair was the way the proof had been achieved. Large and crucial parts of their argument were carried out by a computer, using ideas which had themselves been formulated as a result of computer-based evidence. So great was the amount of computing required that it was not feasible for a human mathematician to check every step. This meant that the whole concept of a 'mathematical proof' had suddenly changed. Something that had been threatening to occur ever since electronic computers were first developed in the early 1950s had finally happened; the computer had taken over from the human mathematician part of the construction of a real mathematical proof.

Until then, a *proof* had been a logically sound piece of reasoning by which one mathematician could convince another of the truth of some assertion. By reading a proof, a mathematician could become convinced of the truth of the statement in question, and also come to understand the reasons for its truth. Indeed, a proof worked as a proof precisely because it did provide those reasons!

Very long proofs such as the classification theorem for simple groups (described in Chapter 5) tend to stretch this simplistic view of a proof to some extent, since the average mathematician, faced with a proof which occupies (say) two 500-page volumes, would be tempted to skip over a great many of the details. But this is really only a matter of economy of effort. Secure in the knowledge that others have checked the various parts of the argument, the busy mathematician need not examine every step in detail. Such proofs are still the product of human endeavour alone. Though computers were used in proving parts of the classification theorem for simple groups, the results they produced could all be checked by hand. The role played by the computer was in no way an 'essential' one.

However, in the proof of the four-colour conjecture the use of the computer was absolutely essential – the proof hinged directly on it. In order to accept the proof you have to believe that the computer program used does what its authors claim of it. When Appel and Haken submitted their proof for publication in the *Illinois Journal of Mathematics*, its editors arranged for the computer part of the proof to be checked *by running an independently produced computer program on another machine*! So a critical part of the proof remained hidden from human eyes.

At first considerable scepticism was voiced by a great many mathematicians. 'Such a procedure, which makes essential use of the results obtained on a computer, results which by their very nature cannot be checked by human hand, cannot be regarded as a mathematical proof,' argued one critic. For such people, the four-colour problem remained unsolved. And indeed, the question of whether or not a 'standard' proof can be found remains open to this day. Given the sheer complexity of the computations involved, even supporters of computer-aided proofs have to concede that the opponents have some justification in their views, and even now at the time of writing, some ten years after the proof was first announced, there are periodic rumours that a subtle error has been found in the computer program, which would render the proof useless. But by and large, with the passage of time and the growing use of computers in society, the number of mathematicians who refuse to accept the proof of the four-colour theorem has gradually declined, and the majority now acknowledge that the arrival of the computer has changed not only the way a lot of mathematical research is carried out, but also the very concept of what is regarded as a proof. Checking the program that produces the 'proof' now has to be allowed as a valid mathematical argument.

And so what was this problem whose solution was to have such a profound effect on the very nature of mathematics? The story begins almost exactly one hundred years before the first commercial computers were built.

Guthrie's Problem

One day in October 1852, shortly after he had completed his studies at University College, London, the young mathematician Francis Guthrie (who was to go on to become Professor of Mathematics at the South African University in Cape Town) was colouring in a map showing the

Figure 31.   Map of the USA. By using four colours it is possible to colour in all the states so that no two states which share a common border are coloured the same. Thus Colorado and New Mexico (for example) have to be coloured differently, though Colorado and Arizona may be coloured the same since they touch only at a point. In a mathematical study states such as Michigan, which consist of two physically separate regions, must be regarded as separate entities. You should have no difficulty in demonstrating for yourself that the map cannot be coloured (in the manner specified) using only three colours.

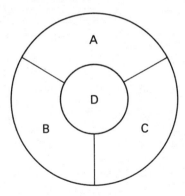

Figure 32. Three colours are not enough. In order to colour the map shown in such a way that no two adjacent countries receive the same colour, you have to colour the countries A, B, C, D using four different colours.

counties of England. As he did so it occurred to him that, in order to colour *any* map (drawn on a plane) subject to the obviously desirable requirement that no two regions (countries, counties, or whatever) sharing a length of common boundary should be given the same colour, the maximum number of colours required seemed likely to be four (see Figure 31). Being unable to prove this, he communicated the problem to his brother Frederick, still at University College as a student of physics. Frederick passed it on to his mathematics professor, the great English mathematician Augustus de Morgan.

Like Francis Guthrie, de Morgan had no difficulty proving that at least four colours are necessary (i.e. that there are maps for which three colours are not sufficient – see Figure 32). He also proved (see later) that it is not possible for five countries to be in a position such that each of them is adjacent to the other four, which at first glance might appear to imply that four colours are always sufficient, but which does not in fact imply this at all (see Figure 33), as de Morgan himself appears to have realized. (Many of the numerous false 'proofs' of the four-colour conjecture that appeared between its formulation in 1852 and its eventual proof in 1976 were based upon a belief in this invalid implication. Indeed, it seems that Francis Guthrie himself at one stage fell into this particular trap.)

Unable to solve the problem, de Morgan passed it on to his students and to other mathematicians (among them Sir William Hamilton of Trinity College, Dublin, the inventor of quaternions – see Chapter 3), giving credit to Guthrie for raising the question. But by and large the problem does not

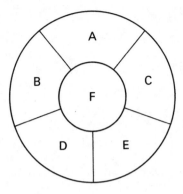

Figure 33. Highlighting a false argument. Many people have assumed that because no map allows the configuration in which each of five countries shares a common border with the other four, four colours will suffice to colour the map. This implication is not valid. In the map shown, there is no configuration in which each of four countries shares a common border with the three adjoining ones, and yet the map as a whole cannot be coloured using three colours, so the number of colours needed to colour a map is *not* the same as the highest number of mutually adjacent countries in the map.

seem to have aroused very much interest until, on 13 June 1878, the English mathematician Arthur Cayley asked the assembled members of the London Mathematical Society if they knew of a proof of the conjecture. (Cayley's question was published in the Society's Proceedings, and this was the first mention of the problem in print.) With this act the hunt was about to begin.

Maps, Networks, and Topology

The first major difficulty facing anyone who sets out to prove the four-colour conjecture is that it refers to *all* maps – not just all the maps in all the atlases around the world, but all conceivable maps, maps with millions (and more) of countries of all shapes and sizes. Knowing that you can colour some particular map using four colours does not help you at all.

You need to produce an argument that will work in all cases. Which means that you have very little to go on. In fact, just what is there to go on? At this stage it is wise to make sure that we are certain what is involved in the problem.

Figure 34. A complicated map that would be difficult (though not impossible!) to colour using just four colours. (Try it!) This particular map was published as part of an April Fool's joke in the magazine *Scientific American* on 1 April 1975. It was included in an article by the celebrated mathematical columnist Martin Gardner who, with his tongue placed firmly in his cheek, claimed that it was an example of a map which refuted the long-standing four-colour conjecture.

For the purposes of Guthrie's problem, a *map* consists of an arbitrary number of regions of the plane ('countries' if you like) separated from each other by lines (or 'borders'). This general definition includes maps of the real world, as in Figure 31, as well as artificial, 'mathematical' maps as in Figures 32, 33, and 34. Actually there is a potential problem with the map of the USA – Figure 31 – in that some states occupy two distinct regions. For instance, Michigan consists of two separate regions on the map, separated by Lake Michigan. As they are physically separate regions they would have to be regarded as such as far as map colourings are concerned. Likewise Long Island, New York, would have to be considered a separate entity from the rest of New York State. Thus as far as a mathematical study of maps is concerned, geometry is the dominant notion, not politics.

The four-colour conjecture concerns the colouring in of the (geographical, not political!) regions of a map in such a way that no two regions which share a common boundary are coloured the same. (Regions which just touch at a point, such as Arizona and Colorado in Figure 31, are not regarded as having a common boundary, and hence may be given the same colour.) What is at issue is the smallest number of colours that you need in order to colour in all the regions of the map in this manner. This is where the second major difficulty lies. Even for a specific map, there are an enormous number of different ways of colouring it in, and it is not the number of colours involved in any one particular colouring that is important, but the least number of colours for which *some* colouring is possible.

If you think about it for a moment or two, you will realize that the actual shapes and sizes of the various regions of the map are not important as far as the colouring problem is concerned – only their positions relative to each other. Thus all the maps shown in Figure 35 are equivalent for the would-be map colourer. The mathematician would express this by saying that what is at issue is the topological structure of the map.

Topology is a mathematical subject much like geometry. In geometry you study properties of objects (or figures) in two, three, or more dimensions ('objects' takes on a highly abstract meaning in four or more dimensions, of course). Likewise in topology. The difference between the two disciplines lies in the type of properties considered. In topology, distance and size are not important, nor is straightness or circularity or angle. In fact topology ignores practically all the properties which are the very life-blood of geometry, studying instead those properties of objects (figures) which remain unchanged under continuous transformations – for example bending, stretching, squashing, or twisting. Two-dimensional topology is sometimes referred to as 'rubber-sheet geometry', since it deals with the properties of figures which would not be changed if the figures were drawn on a 'perfectly elastic' rubber sheet which was then stretched and twisted about (see Figure 36).

To anyone meeting the idea of topology for the first time it would appear that there is not nearly enough to enable a reasonable mathematical study to be made, but in fact nothing could be further from the truth. Topology is a vast area of mathematics in which there are many deep and profound results (see Chapter 10). Indeed, the four-colour problem is itself a problem of topology, though its solution does not use any of the deeper techniques of the subject. Figure 35 illustrates why this is so – the shapes and sizes of the countries that make up a map are not important, only their configuration. You will find it helpful in understanding what follows if you try to bear in mind that it is this *topological* nature of a map that counts, not its superficial 'shape'.

Once this point is grasped, the notion of the *neighbouring network* of a map seems a sensible alternative way of looking at the four-colour problem. Given a map, the neighbouring network is obtained as follows (see Figure 37). Within each region of the map you place a single point, known as a *node* of the network. (You can think of these points as the capital cities

Figure 35. Topological equivalence. Each of the maps shown is equivalent as far as the four-colour problem is concerned; topologically there is no difference between any of them.

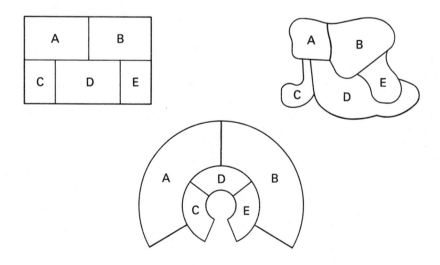

of the countries, if you wish.) You then join up the nodes in a certain way
to form a network (in much the same way that you might link cities by a
rail network). The rule is that two nodes are joined together if, and only if,
their respective map regions share a common boundary, in which case the
line joining them has to lie entirely within the two regions, crossing over
the common boundary. (In terms of a rail link this would mean that the line
cannot cross the territory of any third country.)

The neighbouring network shows at a glance the topological structure of
the map it represents. Indeed, the problem of colouring the map (in the
sense of Guthrie's problem) can be reformulated in terms of colouring the
network: colour the nodes of the network in such a way that any two nodes
which are connected together must have different colours. If all networks
can be so coloured using four colours, so can all maps, and vice versa. So
the network formulation of the four-colour problem provides an alternative
way of looking at it which is entirely equivalent to the original formulation,
and it makes sense to investigate such networks.

This brings the problem into the area known as *graph theory*. Notice that
as a consequence of the way a neighbouring network was defined, no two
paths in the network may cross (or intersect). A *graph* is similar to a

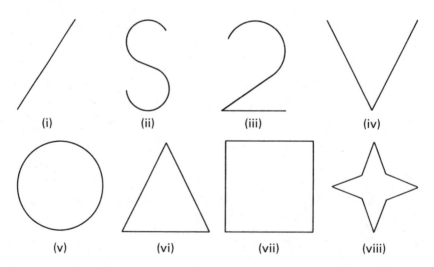

Figure 36.   Topological properties in two dimensions. The objects
(i) to (iv) are topologically the same, and the objects (v) to (viii) are
topologically the same, but none of (i) to (iv) is the same as any of
(v) to (viii).

neighbouring network except that this restriction on paths not crossing is removed. (This use of the word 'graph' is not really connected with its other use in mathematics to refer to curves representing equations drawn on 'graph paper'.) Though much of the original impetus behind the subject of graph theory was provided by the four-colour problem (Hamilton, to whom de Morgan communicated the problem, did a lot of the early work in graph theory), the study of arbitrary 'graphs' is now a large and thriving subject in its own right.

Figure 37. Neighbouring networks. To obtain the neighbouring network for a given map, place a point inside each region of the map and connect together these 'nodes' by lines which lie entirely within the two regions concerned. This is possible only when the two nodes lie in regions which share a common border, in which case the connecting line will cross that border. So the connections reflect the existence of common borders. Colouring the map so that no two adjacent countries are coloured the same is equivalent to colouring the nodes of the neighbouring network so that no two nodes joined by a path of the network are coloured the same. In the example shown it is possible to join each of the nodes with straight lines, but this is not always the case and curved connections are permitted. (Straightness and curving are not topological properties.)

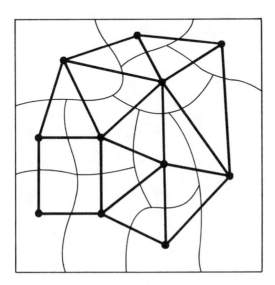

Euler's Formula

A particularly useful observation concerning networks was made (in a slightly different context) by Leonhard Euler. First a little terminology (whose origin will be explained presently). For a start, let us agree that we shall consider only networks which have the property that it is possible to get from any one node to any other by following a route consisting entirely of path connections of the network. This excludes 'pathological' examples of 'networks' such as a set of nodes with no connecting paths, but includes all the networks we shall need for our study of map colouring. Any such network divides the part of the plane which it occupies into a number of regions; these regions are called *faces*. The nodes of the network are sometimes (and in particular in connection with Euler's formula) called *vertices* of the network. The paths which connect these vertices are called *edges*.

At this stage you should draw a number of networks, and for each one tabulate the number (V) of vertices, the number (E) of edges, and the number (F) of faces, as in Figure 38. When you have done this, work out

Figure 38. Euler's formula. For any network, the number (V) of vertices, the number (E) of edges, and the number (F) of faces are such that $V - E + F = 1$.

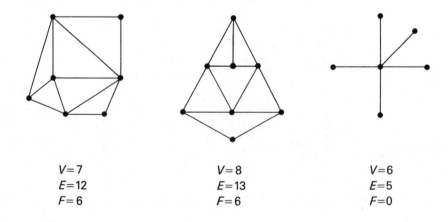

$V=7$	$V=8$	$V=6$
$E=12$	$E=13$	$E=5$
$F=6$	$F=6$	$F=0$

the quantity $V - E + F$ in each case – you will find that the result is always 1. The fact that the equation

$$V - E + F = 1$$

holds for any network was first proved by Euler himself.

In fact Euler was concerned not so much with networks as with polyhedra, which explains the use of the words 'vertex', 'edge', and 'face'. For any polyhedron you find that

$$V - E + F = 2$$

(where V, E, and F have their obvious meanings in terms of a polyhedron). To see that this is essentially the same result as that stated a moment ago for networks, notice that if you remove one face from a polyhedron and then 'stretch out' the remaining figure to lie in a plane, the former edges of the polyhedron will form a network connecting the former vertices (the nodes of the new network), and conversely if you take any network you can 'bend it round' into the shape of a polyhedron with one face missing. It is of course the removed or missing face of the polyhedron that accounts for the difference between the formula for the network and the one for the polyhedron.

Figure 39. Removal of an outside edge from a network decreases both E and F by 1, but leaves V unaltered. This does not affect the value of the quantity $V - E + F$.

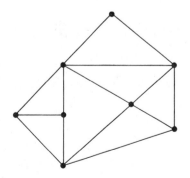
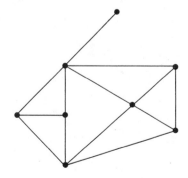

The proof of Euler's network formula provides an excellent example of the kind of argument used in both graph theory and the study of the four-colour problem. Suppose you start with some network and you wish to prove that $V - E + F = 1$. What happens if you remove an outside edge of the network (assuming there is one)? Then E decreases by 1 and so does F, whilst V remains the same (see Figure 39). So the quantity $V - E + F$ remains unaltered by this action. Again, what happens if the network has a 'dangling' vertex (see Figure 40) and you remove both the vertex and the edge leading to it? Then V decreases by 1 and so does E, whilst F is unaltered, so in this case too the quantity $V - E + F$ is unchanged. Now, if you start with your given network and, like the sea eroding an island, keep on removing outside edges and dangling vertices whenever possible, you will eventually end up with just a single vertex. That is, you will have reduced your original network to a trivial one in which $V = 1$, $E = 0$, and $F = 0$. In this final 'network' the quantity $V - E + F$ is equal to 1. But none of the 'erosion' operations you performed altered $V - E + F$. So the value of this expression *in the original network* must have been 1 as well. And that proves the result. If you wish, you can try this out for yourself. Start with some arbitrarily drawn network and keep on removing outside edges and dangling vertices, tabulating the values of V, E, F, and $V - E + F$ as you go.

Figure 40. Removal of a 'dangling' vertex from a network decreases both V and E by 1, but leaves F unaltered. This does not affect the value of the quantity $V - E + F$.

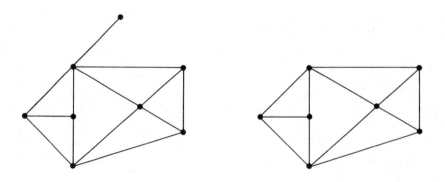

de Morgan's Theorem

The one positive result that de Morgan managed to obtain on Guthrie's problem was to prove that in no map can the situation occur where each of five countries borders the other four. By using the neighbouring network together with Euler's formula, this is quite easy to demonstrate. In terms of networks, what de Morgan's result amounts to is that it is not possible to draw a network with five vertices so that each vertex is connected to the other four. Certainly if you try to do this (see Figure 41) you will invariably find that you are left with two vertices which cannot be connected without crossing over a path that has already been drawn, but this does not constitute a proof, since it may just be that you have drawn the earlier connections inappropriately. The following argument does not depend on a particular drawing, and thus provides a rigorous proof.

Figure 41. It is not possible to draw a network with five vertices such that each vertex is connected to the other four. No matter how you try to join up the vertices, you will be left with two vertices (A and E in the network shown) that cannot be joined without crossing one of the lines already drawn.

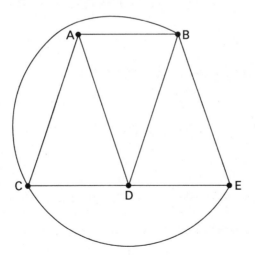

Suppose that it were possible to draw a network with five vertices such that each vertex is connected to all the others. If the area surrounding the network is regarded as an additional 'face', then each edge of the network will separate two faces. Moreover, since there is now one extra face, Euler's formula becomes

$$V - E + F = 2.$$

We know the value of V here, it is 5. Also, since every vertex is joined by an edge to every other, $E = 10$. (Check this for yourself!) So F has to be 7 if the above Euler formula is to hold.

So far so good. Now we perform another calculation. Since each face will be surrounded by at least three edges (this is true also for the new 'face' we introduced a moment ago, though you have to understand the word 'surrounded' is used in a topological sense in this case), counting edges by faces gives at least $3 \times 7 = 21$ edges. But if you count edges in this way (by faces), each edge will be counted twice, since it separates two faces. So the correct answer is that there must be at least $\frac{1}{2} \times 21 = 10\frac{1}{2}$ edges, which (since there is no such thing as half an edge) means that there are at least 11 edges. But, as we noted above, $E = 10$. So we have arrived at a contradictory situation, and as usual in such arguments the conclusion is that the original assumption of the argument must be false – that is, it is *not* possible to draw a network with five vertices so that each vertex is connected to all the others. That proves de Morgan's theorem about maps.

The Five-Colour Theorem

In 1879, within a year of Cayley presenting the four-colour problem to the London Mathematical Society, one of its members, a barrister called Alfred Bray Kempe, published a paper in which he claimed to prove the conjecture. But he was mistaken, and eleven years later Percy John Heawood pointed out a significant error in the argument. Heawood was however able to salvage enough to prove that five colours are always adequate, and the proof of this 'five-colour theorem' is sufficiently simple to give here.

First of all, notice that by reasoning as we did above to relate the Euler formula for networks to that for polyhedra, we may conclude that it does not matter (as far as the four- or the five-colour theorem is concerned) whether we draw our maps on the plane or on the surface of a sphere. If we start with a map on a sphere we can deform it to an equivalent map on the plane by piercing a hole in the middle of one of the regions and pulling the entire map out flat (so that the pierced region becomes one which surrounds the rest of the map). Conversely, if we are given a map on the plane we may regard the region surrounding the map as an extra country, and fold the entire map round into the shape of a sphere (bringing the added surrounding region together to form an 'enclosed' region just like all the others). This procedure shows that if every planar map can be coloured with N colours, so can every map on the sphere, and vice versa.

We shall in fact prove the 'five-colour theorem' for maps drawn on a sphere. We shall make use of the Euler formula, which for maps on a sphere is

$$V - E + F = 2.$$

Our use of this formula will be in connection with the map itself, rather than with the associated neighbouring network as was the case with the proof of de Morgan's theorem. (So a *face* will be a region of the map, an *edge* will be a border, and a *vertex* will just be a point where three or more borders meet.)

The idea of the proof is to start with a given (entirely arbitrary) map drawn on a sphere, and gradually modify it by a process of merging two or more adjacent countries into one, so that eventually a map is obtained having at most five countries – which can obviously be coloured using five colours or fewer. Provided the steps used in the modification process do not reduce the number of colours required to colour the map, this will prove that five colours suffice for the original map. So the crux of the proof is to describe the individual processes which are used to *reduce* a given map to a simpler one (i.e. one with fewer countries) without reducing the number of colours necessary to colour the map. There are six different *reduction processes*, each of which is applicable in a different situation depending on particular configurations of countries on the map.

First of all, if one region is entirely surrounded by another (see Figure 42(i)), then the inner region may be merged with the surrounding one. Any colouring of the new map using at least two colours can be extended to a colouring of the original map using the same colours: the inner region is simply assigned a colour other than the one used to colour the entire merged region on the modified map.

The next reduction operation applies whenever there is a vertex at which more than three regions touch. For if at least four regions touch, then (and you may need to think about this for a moment) one pair of these regions will not have a common border (anywhere on the map!) and these two regions can be merged into one (see Figure 42(ii)). Given any colouring of the modified map, the original map may be coloured using the same number of colours by assigning the same colour to the two regions that were merged, and colouring the rest of the map the same in both cases. By applying this reduction repeatedly the map can be modified so that only three regions touch at each vertex. This will be assumed to be the case for the remainder of the reductions.

If there is a region which borders on just two others (see Figure 42(iii)), then that region may be 'removed' by merging it with one of these two. If the new map can be coloured using at least three colours, the original map can be coloured using the same colours simply by colouring the merged central region differently from the two surrounding areas.

Any region having three neighbours can be 'removed' by merging it with one of its neighbours (see Figure 42(iv)), and as in the previous case if the new map can be coloured using at least four colours then the original can be coloured using the same colours.

Likewise any region having four neighbours can be merged with one of its neighbours (see Figure 42(v)), and this will not involve any change in the colour requirements when five colours are available.

By applying the above reduction procedures as often as possible, you will end up with a map in which no region is surrounded by another, in which each vertex lies on exactly three edges, and all of whose regions have at least five edges. In fact, at least one region will have exactly five edges, as we now prove.

There are V vertices, E edges, and F regions. Let a be the average of the number of edges bordering each region. (So a might be fractional.) Since each edge lies between two regions,

$$2E = aF.$$

Figure 42 (facing).   Reductions used in the proof of the five-colour theorem (see the text for details).

(i)

(ii)

(iii)

(iv)

(v)

(vi)

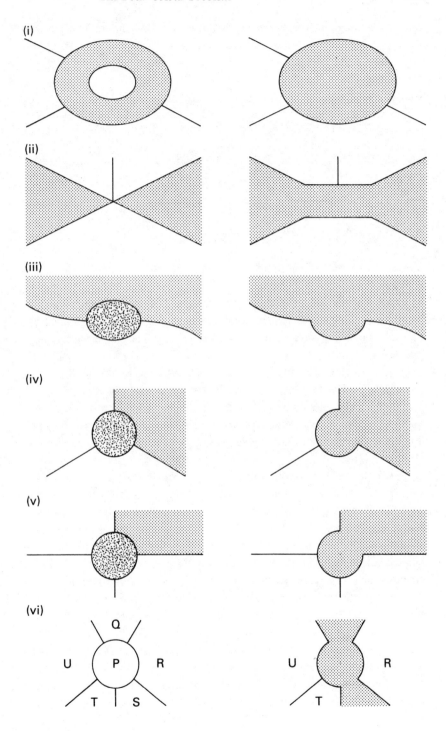

Also, each vertex lies on three edges and each edge joins two vertices, so

$$3V \; = \; 2E.$$

Hence

$$3V \; = \; 2E \; = \; aF.$$

Substituting $V = \frac{1}{3}aF$ and $E = \frac{1}{2}aF$ in Euler's formula $V - E + F = 2$ gives

$$\tfrac{1}{3}aF - \tfrac{1}{2}aF + F \; = \; 2,$$

so

$$a \; = \; 6 - \frac{12}{F}.$$

Thus a is less than 6. Since the average number of edges for each region is less than 6, some regions must have less than 6 edges. But all regions have at least 5 edges. So some regions will have exactly 5 edges, as we set out to prove.

Now consider such a five-edged region P, with neighbours Q, R, S, T, U, as in Figure 42(vi). One pair of neighbours of P do not touch, say Q and S. Merge the three regions P, Q, S. If the new map can be coloured with five colours, so can the original. In the merged map Q and S have the same colour, so there are four colours surrounding P, which leaves one to spare for P.

That completes the reduction procedure. Since each step reduces the number of regions on the map, by applying it repeatedly you will eventually arrive at a map with five or fewer regions. Since any such map can obviously be coloured with five colours, so can the original map. Indeed, by working back through the various reductions a colouring of the map using five colours can actually be built up in an entirely mechanical way. (Try it out for yourself, starting with a moderately complicated-looking map.)

The basic idea of reducing the map step by step, as in the above proof, has been used in practically every serious attempt to prove the four-colour conjecture (including the final successful attempt) since Kempe first introduced the device in his flawed attempt to solve the problem. Since the actual solution was essentially an extension of what Kempe did (though a quite considerable extension), it is worth while taking a closer look at his argument.

Kempe's Method

The proof of the five-colour theorem given above, whilst salvaged from Kempe's false argument by Heawood, is much simpler than the one Kempe thought adequate for the four-colour theorem. (Obviously the arguments used in the reductions illustrated in Figure 42(v) and (vi) do not work when only four colours are available.) Kempe's procedure was roughly as follows.

Start off by assuming the existence of a map for which five colours are required (i.e. four colours are definitely not sufficient), then argue to obtain a contradiction. The first step is to define the notion of a *normal map*. This is one such that no country is entirely surrounded by any other and no more than three countries meet at any point. By applying the first two of the reduction procedures described above, starting from a map which requires five colours, you can obtain a normal map requiring five colours. Since the existence of a map requiring five colours is assumed, it follows that there will be a normal map which requires five colours. But of course there may be many such, with different numbers of countries. Amongst these there will be at least one having the smallest number of countries (for such a map). Kempe tried to obtain his contradiction by working with such a *minimal normal map* requiring five colours.

The point about using a minimal map is that any (normal) map with fewer regions may be coloured using four colours, so if you can find a reduction operation which will reduce the size of your map by even one country, without altering the requirement for five colours, you will at once have your contradiction since the reduced map cannot simultaneously be colourable using four colours and not colourable with fewer than five colours.

Kempe proved (correctly) that in any normal map there has to be some country with at most five neighbours, i.e. at least one of the configurations shown in Figure 42(iii), (iv), (v), and (vi) has to occur somewhere in the map. He then argued (incorrectly) that if a minimal normal map requiring five colours has a country with at most five neighbours, then it could be reduced to a normal map with fewer countries which still required five colours, thereby arriving at a contradiction as outlined a moment ago. His arguments were perfectly correct as far as countries with two, three, or four neighbours are concerned. The arguments for the two- and three-neighbour cases were given above in the proof of the five-colour theorem. But if there are four neighbours we need a different and more subtle argument. It is necessary to examine the part of the map which surrounds the configuration, and possibly alter the colours of some of the surrounding countries. It is not unduly difficult, but it does require some considerable thought and ingenuity. Where Kempe went wrong was – as Heawood noticed – in the case of a country with five neighbours (see Figure 42(vi)).

Nevertheless, Kempe's argument already involved two of the key ideas that would eventually be used in the solution to the problem. One is the notion of an *unavoidable set* of configurations – a collection of map configurations such that any minimal normal map requiring five colours has to contain at least one of them. (Kempe's unavoidable set consisted of the configurations shown in Figure 42(iii), (iv), (v), and (vi).) The other is the notion of *reducibility*: if a certain configuration appears in a minimal normal map requiring five colours, then it is possible to reduce the number of countries in the map so as to produce the contradictory situation of a normal map requiring five colours having fewer countries than the minimal one. Provided you can demonstrate that every configuration in an unavoidable set *is* reducible, you can prove the four-colour theorem. Kempe's argument failed because his proof of reducibility did not work for one of the four configurations in his unavoidable set. The Appel–Haken proof succeeded by making a detailed analysis of this flawed last case, and this required the discovery of a different unavoidable set. You can begin to get some idea of the difficulty that faced them when you consider that the unavoidable set that they eventually constructed contained almost 1500 configurations. How such a set was discovered will be explained in due course. But first, what progress was made on the problem in the years between Kempe's false proof and the Appel–Haken solution?

Heawood's Formula

The work by Heawood has already been mentioned. After spotting the flaw in Kempe's argument in 1890, he went on to spend the next 60 years of his life working on the problem, and though he did not

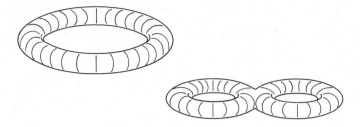

Figure 43. A torus and a double-torus.

solve it he did succeed in resolving the analogous question for maps drawn on surfaces other than planes or spheres. His solution involves the so-called *Euler characteristic* of a surface. If you take any surface, say a sphere, a torus, or even a double-torus (see Figure 43), and draw a map covering the entire surface, the quantity $V - E + F$ will work out to be the same however you draw the map – just as it does for maps drawn on a sphere (for which the answer is 2). This quantity, which therefore depends not on the map (since any map is as good as any other here) but on the surface (different surfaces give different answers), is called the *Euler characteristic* of the surface. It is a *topological invariant* of the surface, which is to say it remains the same no matter how the surface is topologically manipulated. For the torus the Euler characteristic is 0; for the double-torus it is -2. For the *Klein bottle*, a curious surface which has no edges and only one side, and which can be constructed in three-dimensional space only if you allow self-intersection (see Figure 44), the Euler characteristic is 0, the same as for the torus. But note that although they have the same Euler characteristic, the torus and the Klein bottle are *not* topologically equivalent: you cannot deform one into the other. Though topologically equivalent surfaces *do* have the same Euler characteristic, topologically different surfaces may or may not.

By employing essentially the same kind of argument used to prove the five-colour theorem for maps drawn on a sphere, Heawood proved that for a surface of Euler characteristic n,

$$\tfrac{1}{2}(7 + \sqrt{49 - 24n})$$

colours suffice to colour all maps drawn on the surface, provided n is at most equal to 1. Unfortunately, the only surface for which n is greater than 1 is the sphere, for which $n = 2$, and so Heawood's otherwise splendid result

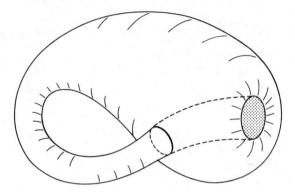

Figure 44. The Klein bottle – a topological figure that consists of a surface having no edge and only one side. In three-dimensional space this can only be achieved if the surface is allowed to 'pass through' itself. In four dimensions such self-intersection would not be necessary.

just failed to handle the one case everyone was most interested in – for if you put $n = 2$ in Heawood's formula, you get the answer 4, of course!

Thus for the torus, for which $n = 0$, seven colours are sufficient, and since it is not hard to draw a map on the torus which cannot be coloured with six colours, this proves the 'seven-colour theorem' for maps drawn on a torus in the strong sense that seven colours suffice and fewer colours do not. In fact, in 1968 it was shown by Ringel and Youngs that Heawood's formula gives the exact minimal number of colours required in all cases except for the sphere (for which at that time the answer was not known) and the Klein bottle (for which $n = 0$ and the formula gives the answer 7, though another argument shows that only six colours are needed*).

Towards a Four-Colour Theorem

After Heawood's work, numerous mathematicians (and an even greater number of amateurs) investigated the four-colour problem, developing in the process a great many mathematical techniques that

---

*So, as a consequence of the four-colour theorem it is now known that the Klein bottle is the *only* surface for which Heawood's formula does not give the exact, minimal answer.

ultimately proved to have applications elsewhere in mathematics. Of all this effort, some of it can, with the benefit of hindsight, be seen to have led towards the final solution to the problem. Briefly, this is what happened.

In 1913 George Birkhoff improved on Kempe's reduction technique and managed to show that certain configurations larger than Kempe's were reducible. In 1922 Franklin used some of Birkhoff's results to show that every map with 25 or fewer countries could be coloured with four colours. In 1926, Reynolds improved this result to 27 countries, and then in 1938 Franklin regained the 'record' by going up to 31. Winn, in 1940, reached 35, and there progress stopped until 1970, when Ore and Stemple proved the four-colour theorem for all maps having fewer than 40 countries. The figure reached 96 before the Appel–Haken proof finally rendered all such results superfluous.

But although all of this work did show that many configurations are reducible, the set of all configurations that had been proved reducible by 1970 did not even come close to forming an unavoidable set (as would be required to prove the four-colour conjecture). Various unavoidable sets had been constructed, but none seemed likely to lead to an unavoidable set of reducible configurations. Either you had reducibility or you had an unavoidable set, but not both. In 1950, the German mathematician Heinrich Heesch, who had been working on the four-colour problem since 1936, estimated that an unavoidable set of reducible configurations would have to contain about 10 000 separate configurations. Though this estimate was ultimately to prove grossly overgenerous, it was accurate in indicating that the problem would be solved only with the aid of very powerful computational equipment capable of handling vast amounts of data. Indeed, Heesch himself, realizing that the ability to handle large sets of configurations was likely to be the key to the solution, was the first to advocate – and attempt – a computer-aided assault on the problem.

He began by formalizing the several known methods of proving configurations reducible, and noted that at least one of them (a straightforward generalization of Kempe's method) was sufficiently mechanical to be implemented on a computer. Karl Dürre, a student of Heesch, then wrote a program to prove reducibility. All this was done in terms of the neighbouring network representation of the map, which provided a more convenient way of handling the problem on a computer.

One problem that had to be overcome was that the failure of one method of proving that a particular configuration was reducible did not necessarily mean that the configuration could not be reduced – another method might succeed where the first had failed. To deal with this difficulty it was necessary to develop what might be called a small 'arsenal' of techniques for proving reducibility. By the late 1960s, Heesch had built up a large enough arsenal for Appel and Haken to use at the start of their final assault on the problem in 1976.

However, comparable progress had not been made with the construction of an unavoidable set of configurations. Heesch did try a method that was analogous to moving a charge around an electrical network, but he did not pursue it very far. Perhaps he should have. For this was the trick that was to lead to the final solution.

Heesch's Charge Method

The neighbouring network associated with a minimal normal map requiring five colours is (and this follows from Kempe's work) one in which every face is a triangle and at least five edges meet at each vertex. (The number of edges meeting at a vertex is called the *degree* of that vertex.) The idea is now to regard the network as an electrical one and assign a charge to each vertex. If a vertex has degree k, it is given charge $6 - k$. Thus vertices of degree 5 have a positive charge (of $+1$), vertices of degree 6 have no charge, vertices of degree 7 have charge -1, and so on. It follows from Kempe's work that the sum of the charges over the entire network is (for any network) 12. The precise value of 12 will not be important; what is important is that the sum of the charges is always positive.

Now suppose you start to move positive charge around the network (in fractional amounts if you wish). This will, of course, not cause any net gain or loss in the total charge in the network, but some degree-5 vertices may end up losing all of their charge (i.e. become *discharged*), whilst some vertices of degree greater than 6 may end up with a positive charge (i.e. become *charged*). The exact final situation will obviously depend on the precise *redistribution* (or *discharging*) procedure used. But (and this is the key), because it is possible to determine the layout of small pieces of a map without knowing the entire map, then given a specified discharging procedure (applicable to any map) it is possible to generate a finite list of all the configurations that will end up with net positive charge.

Now, because the total charge on the network is positive, there will always be some vertices with positive charge. So, as all possible receivers of

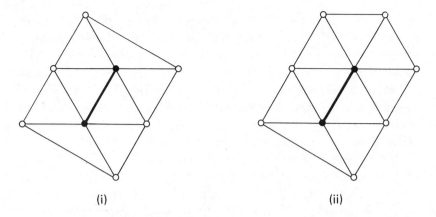

(i) (ii)

Figure 45. An unavoidable set generated by the simple discharging procedure described in the text. The unavoidable set consists of the configurations (i) and (ii). In (i) the configuration is produced by two degree-5 vertices joined together, in (ii) by a degree-5 vertex joined to a degree-6 vertex. These two vertex pairs are shown as black circles connected by a heavy line. The remainder of each of the two configurations is then determined by the degrees of the vertices in the producing pair and the fact that the network is triangular. (There is no restriction on the degree of each of the outer vertices, shown as white circles.) Unavoidability means that at least one of the two networks (i) and (ii) will always be found.

positive charge are included in the finite list of configurations that has been generated from the discharging procedure, every network (of the kind now being considered) must contain at least one of these configurations. In other words, the list of configurations generated will form an unavoidable set, which is what we are looking for. Kempe's original unavoidable set can be regarded as the one resulting from the trivial procedure of moving no charge at all. Thus the discharging procedure is a generalization of Kempe's method. And being more general, it should have a greater chance of success!

A simple example should help to make things clear (though it is likely that you will need to 'play around' with this example for some time in order to see what is going on). Suppose your discharging procedure consists of transferring $\frac{1}{5}$ of a unit of charge from every degree-5 vertex to each neighbouring vertex having degree 7 or more. Then the resulting unavoidable set consists of just the two configurations shown in Figure 45. To see how these arise, note first that a degree-5 vertex can end up positive only if it has at least one neighbour of degree 5 (Figure 45(i)) or degree 6 (Figure 45(ii)). A degree-6 vertex starts out with no charge, and does not receive any under this procedure. A degree-7 vertex can become positive only if it has at least six degree-5 neighbours; if this is the case then, because every face is a triangle, two of these neighbours are joined by an edge (so Figure 45(i) applies to that pair of neighbours). A vertex of degree 8 or more cannot become positive even if all its neighbours have degree 5. Moving $\frac{1}{5}$ of a unit simply will not be enough. (For example, for a degree-8 vertex with eight degree-5 neighbours, the original charge will be -2 and $8 \times \frac{1}{5} = 1\frac{3}{5}$ units will be transferred, leaving a final charge of $-\frac{2}{5}$.)

Thus the two configurations shown form an unavoidable set. That is, since the above argument works for any network (of the type being considered), at least one of these two configurations will always be found.

The idea behind the use of the charge method to prove the four-colour conjecture is to find a discharging procedure such that the resulting unavoidable set consists entirely of reducible configurations. If this can be done, the theorem follows at once. (Neither of the two configurations produced in the example above is reducible, by the way.)

Proof of the Four-Colour Theorem

It was in 1970 that Wolfgang Haken hit upon some new methods for improving discharging procedures, and though the task seemed a formidable one that would require immense amounts of computation time, he began to hope that this might lead eventually to a proof of the four-colour conjecture. In 1972, he and Kenneth Appel set to work in earnest to try to bring this hope to fruition.

Their goal was to devise a discharging procedure that would lead to an unavoidable set of reducible configurations. This involved two things: finding the discharging procedure and proving the reducibility of the unavoidable configurations it gave rise to. At first they worked with highly restricted types of network which previous work (by Heesch and others) had shown ought to be easier to deal with. The general strategy was clear: start with a promising-looking discharging procedure and try to prove that each of the resulting unavoidable configurations is reducible. If reducibility could not be proved for one or more configurations in the list, amend the discharging procedure so that the troublesome configuration(s) no longer appeared. Though simple to describe, carrying out this strategy was not at all an easy matter, and involved many weeks of man–machine 'dialogue' as first one then another discharging procedure was tried, but very gradually progress appeared to be made. A similar approach of computer 'experiment' punctuated by human intervention was adopted as the pair carried out a simultaneous search for improved methods of proving reducibility. After three years of this, by the beginning of 1976, they finally felt they had enough information to begin their assault on the full problem. As a result of all their experimental work they had developed a discharging procedure which looked likely to produce an unavoidable set of reducible configurations, and had written a routine for proving reducibility which looked like working on the kinds of configuration they would encounter. Their computer program had a self-modifying feature so that if it encountered a configuration that could not be proved reducible, positive charge would be moved around in order to try to avoid the difficulty. But would it all work? Could it all work? The only way to find out was to start the program running and see what happened. And that is what they did.

Six months later, in June 1976, they had their answer. Their program had (with considerable help from its two by now very expert creators) succeeded in proving the four-colour theorem. It had taken four years of intense work and 1200 hours of computer time. The discharging procedure that finally did the trick differed from the one they had started with by some 500 modifications, suggested by the results of the various computer runs. The two mathematicians themselves had to analyse some 10 000 neighbourhoods of positively charged vertices *by hand calculation*, and the computer had to examine over 2000 configurations and prove the reducibility of a total of 1482 configurations in the unavoidable set. But it all worked. A hundred years of effort had finally come to an end.

Mathematics would never be quite the same again.

Suggested Further Reading

A readable and fairly comprehensive account of the solution to the four-colour problem can be found in the article 'The solution of the four-color-map problem', written by Kenneth Appel and Wolfgang Haken for the magazine *Scientific American*, Volume 237 (October 1977), pp. 108–21.

In response to a number of criticisms that their proof was flawed, the same authors wrote another account of the proof (slightly more mathematical than the one referred to above, but still at an expository level) for the magazine *The Mathematical Intelligencer*, Volume 8 (1986), pp. 10–20. The title of the article is 'The four color proof suffices', a variant on the special postmark that the local Illinois post office used to honour the solution to the problem in 1976, which announced proudly 'four colors suffice', no doubt much to the bafflement of the uninitiated who chanced to receive a letter bearing this message.

For a detailed, mathematical coverage of both the problem and its solution there is the book *The Four Colour Problem*, by Thomas L. Saaty and Paul C. Kainen (McGraw-Hill, 1977).

A nice, comprehensive coverage of the development of graph theory in general is provided by the book *Graph Theory 1736–1936*, by N. Biggs, E. Lloyd, and R. Wilson (Oxford University Press, 1976).

8 Fermat's Last Theorem

The Most Famous Problem in Mathematics

Early in 1983, a 29-year-old West German mathematician, Gerd Faltings, proved a result which marked the first major progress for over 100 years on what is undoubtedly the most famous unsolved problem in mathematics. That problem is (of course) *Fermat's Last Theorem*, a 300-year-old teaser whose fame is not restricted to the mathematical world. Indeed, there can scarcely be an educated person anywhere who has not heard of it – even if they do not know what it says. And yet the origin of the problem amounts to no more than a scribbled note in the margin of a book.

When Pierre de Fermat died on 12 January 1665, he was one of the most famous mathematicians in Europe. Though his name is nowadays invariably associated with number theory, much of his work in this area was so far ahead of his time that to his contemporaries he was better known for his research in coordinate geometry (which he invented independently of Descartes), infinitesimal calculus (which Newton and Leibniz brought to fruition), and probability theory (which was essentially founded by Fermat and Pascal). And yet for all that he was not a mathematician by profession, but a lawyer and magistrate attached to the provincial parliament in Toulouse, a position which he attained in 1631 when he was 30 years old.

It was after he had taken up his jurist's post that he began to devote his leisure time to mathematics – a subject in which he had had no formal training, but for which he rapidly developed a great affinity. What he did not develop an affinity for was the preparation of his work for publication. With only minor exceptions, he published virtually nothing throughout his mathematical career. But he did keep up a copious correspondence with his contemporaries – contemporaries who were amongst the greatest mathematicians alive. In a world peopled by such giants of mathematics as Desargues, Descartes, Pascal, Wallis, and Jacques Bernoulli, this Frenchman for whom mathematics was a hobby could count himself the equal of any: Pierre de Fermat, the 'Prince of Amateurs'.

The path leading to the formulation of the famous Last Theorem is a long and interesting one. When Constantinople fell to the Turks in 1453 the Byzantine scholars fled to the West, bringing with them the ancient manuscripts of Greek learning. Among them was a copy of what had survived of Diophantus' *Arithmetica*. This work remained safe but largely unread until 1621, when Claude Bachet published a new edition of the original Greek text, along with a Latin translation containing notes and comments. This brought the book to the attention of the European mathematicians, and it seems that it was through reading the *Arithmetica* that Fermat first became interested in number theory.

The *Arithmetica*, written sometime in the third century A D, was Diophantus' principal work, and was one of the first books on algebra to be written. The greater part of the treatise concerns the solution in rational numbers of equations in two or more variables having integer coefficients. Present-day mathematicians, when working on such problems, usually restrict themselves to finding integer solutions. This often amounts to the same thing. For example, for a linear equation in three variables such as

$$2X + 3Y + 4Z = 0,$$

the rational solution $X = \frac{1}{4}$, $Y = \frac{1}{10}$, $Z = -\frac{1}{5}$ can be converted to the integer solution $X = 5, Y = 2, Z = -4$ by multiplying through by 20, the least common multiple of 4, 10, and 5. A similar procedure can be used in many other cases to convert a rational solution into one consisting entirely of integers. Certainly this is true of all the equations to be considered in this chapter, and so for the most part we too will consider only integer solutions.

As he worked through his copy of the Bachet edition of the *Arithmetica*, Fermat was in the habit of making brief notes in the margin. When, five years after Fermat's death, his son Samuel set about collecting together all of his father's notes and letters for publication, he came across the annotated copy of the *Arithmetica* and decided to publish a new edition of the book, including Fermat's marginal notes as an appendix. The second of these 48 *Observations on Diophantus* (as Samuel called them) had been written by Fermat in the margin next to Diophantus' Problem 8 in Book II, which asks 'Given a number which is a square, write it as a sum of two other squares.' Fermat's note said (in Latin) that:

> On the other hand, it is impossible for a cube to be written as a sum of two cubes or a fourth power to be written as a sum of two fourth powers or, in general, for any number which is a power greater than the second to be written as a sum of two like powers. I have a truly marvellous demonstration of this proposition which this margin is too narrow to contain.

In algebraic terms, Diophantus' problem asks for rational numbers x, y, z satisfying the equation

$$x^2 + y^2 = z^2.$$

This turns out to be a fairly easy task. What Fermat's marginal comment asserts is that if n is a natural number greater than 2, then the equation

$$x^n + y^n = z^n$$

has *no* rational solutions. (As mentioned in Chapter 3, in Diophantus' time – and indeed to some extent in Fermat's time – 0 was not considered to be a number, so the trivial solutions you get by putting one of the variables equal to 0 are excluded here. The problem concerns positive rational solutions only.)

Note that, by means of a simple argument as above, it does not make any real difference if we amend both Diophantus' problem and Fermat's claim to refer to integer solutions (in fact positive integer solutions) rather than rational solutions, since any rational solution will lead at once to an integer solution (and conversely any integer solution is clearly a rational solution). Thus we can take Fermat's Last Theorem (as his marginal claim is called) to be the assertion that, for any natural number n greater than 2, the equation

$$x^n + y^n = z^n$$

has no positive integer solutions.

But why is it called 'Fermat's Last Theorem'? The origin of this name is somewhat obscure. Though it is not known for certain when Fermat made his famous marginal note, it seems likely that it was during the period when he was first studying Diophantus' book, sometime in the 1630s. But this was at the start of his mathematical career, so the theorem was surely not his last one. Much more likely the name stems from the fact that, of all the many statements of theorems he left behind after his death, this is the last one that remains to be proved (if proved it can be).

That might explain the use of the word 'last', but what of 'theorem'? Did Fermat really have the 'truly marvellous demonstration' that he claimed? Though this has to be allowed as a possibility, the evidence suggests that he was mistaken – and indeed that he himself later realized his error. His other theorems are stated and restated in letters and challenge problems to other mathematicians, and the two special cases $x^3 + y^3 = z^3$ and $x^4 + y^4 = z^4$ of the Last Theorem are also stated elsewhere, whereas the only mention of the full Last Theorem is that one brief marginal note. Very likely he saw how to prove it for $n = 4$, and possibly also for $n = 3$ (for which two exponents the theorem is now definitely known to be true), and thought his arguments could be generalized to cover every other integer n, but subsequently discovered that this was not so. Since his marginal notes were never intended for publication, he would have had no need to go back and make any alteration. Indeed, he may well have forgotten all about his earlier entry!

And yet in the popular mind the belief persists that Fermat really did have a proof. It is, after all, a wonderful story: a seventeenth-century amateur proving a result that was to defeat the efforts of professional mathematicians for the next 350 years (at least). The fact that the problem is so easy to state makes the story even better, of course. And then there is always the possibility that Fermat *was* right!

But whether Fermat had a proof or not, the fact remains that no one else has been able to solve this tantalizingly simple-looking problem one way or the other. And this is not for want of trying. Many great mathematicians have spent years of their lives wrestling with it, and work on the problem has led to totally new areas of mathematics being developed (see later). Entire books have been written about it (some of which are listed at the end of this chapter). Indeed, the results that have been obtained as a consequence of trying to prove the Last Theorem now far outweigh the theorem itself in their significance for the rest of mathematics. If Fermat's Last Theorem were proved tomorrow, virtually no new mathematical results would follow from it. Its importance rests solely on two things: its fame and the very fact that no one has been able to solve it!

So just what is known about the Last Theorem, and what was the 'major advance' made by that German mathematician in 1983? This chapter sets out to answer those questions.

Pythagorean Triples

The problem in Diophantus' *Arithmetica* which led to the formulation of Fermat's Last Theorem is to find a method for the solution (amongst the rational numbers, though we shall restrict ourselves to integer solutions) of the equation

$$x^2 + y^2 = z^2.$$

Because of the obvious connection with Pythagoras' theorem, any three integers x, y, z which satisfy the above equation are said to form a *Pythagorean triple*. For example, the numbers 3, 4, 5 form a Pythagorean triple because

$$3^2 + 4^2 = 5^2.$$

Once you have a Pythagorean triple, you can obtain from it an infinite collection of other Pythagorean triples. Simply multiply the three numbers in your triple by any other number you like. For example, multiplying the triple 3, 4, 5 by 2 gives 6, 8, 10, and this is a Pythagorean triple because

$$6^2 + 8^2 = 10^2.$$

Multiplying by 3 gives the Pythagorean triple 9, 12, 15. And so on. But there is a very real sense in which there is only one *solution* involved here, namely 3, 4, 5, with the others simply 'variations' on it. The solution 5, 12, 13, on the other hand, is a quite different solution (which in turn will give rise to its own infinite family of solutions). What distinguishes the solutions 3, 4, 5 and 5, 12, 13 from the infinitudes of solutions that arise from them by multiplication by a constant is that these original solutions have no common factors: there is no number which divides each of 3, 4, and 5, and no number which divides 5, 12, and 13.

In general, if *a*, *b*, *c* is any Pythagorean triple, then so is any multiple *ma*, *mb*, *mc*; and conversely, if *u*, *v*, *w* is any Pythagorean triple and if *d* is a common factor of *u*, *v*, and *w*, then *u*/*d*, *v*/*d*, *w*/*d* is also a Pythagorean triple. In order to highlight the special nature of the 'basic' triples such as 3, 4, 5 and 5, 12, 13, mathematicians call Pythagorean triples *x*, *y*, *z* which have no common factor (other than 1) *primitive* Pythagorean triples. Essentially, then, Diophantus' problem is concerned with finding a way of determining all primitive Pythagorean triples.

A fairly straightforward piece of mathematical reasoning leads to the following formula for generating all possible primitive Pythagorean triples *x*, *y*, *z*:

$$x \; = \; 2st, \quad y \; = \; s^2 - t^2, \quad z \; = \; s^2 + t^2,$$

where *s* and *t* are any natural numbers such that *s* is greater than *t*, *s* and *t* have no common factor, and one of *s* and *t* is even, the other odd. So, for example, $s = 2, t = 1$ gives the triple $x = 4, y = 3, z = 5$; $s = 3, t = 2$ gives $x = 12, y = 5, z = 13$; $s = 4, t = 1$ gives $x = 8, y = 15, z = 17$; and so on.

The above complete solution to Diophantus' problem appeared in Euclid's *Elements* (*c.* 350–300 BC).

The Case *n* = 4

Having disposed of Diophantus' problem, what of Fermat's Last Theorem itself? This asserts (in integer form) that for every natural number *n* greater than 2, the equation

$$x^n + y^n \; = \; z^n$$

has no (positive) integer solutions. Just how do you go about proving (or rather trying to prove) this kind of statement?

A sensible first step is to look at a few special cases, say $n = 3, n = 4$, and $n = 5$; if you can solve those, you might see how to prove the entire theorem. This seems to be how Fermat himself must have approached the matter. The only concrete evidence we have concerns his work on a problem closely related to the case $n = 4$. Indeed, this work is practically the only piece of Fermat's own mathematical reasoning that anyone besides himself has ever seen! It consists of yet another marginal note in the *Arithmetica*. Interestingly enough in view of the marginal note in which the Last Theorem was stated, this particular note ends with the words: 'The margin is too small to enable me to give the proof completely and with all detail.'

Before I explain Fermat's argument (and how it yields the case $n = 4$ of the Last Theorem), you might like to ask yourself just how, in general terms, you would go about trying to deal with (say) the $n = 4$ problem. You could start off by trying a few values for x, y, z to see if any of them satisfy the equation involved, namely

$$x^4 + y^4 = z^4.$$

(Your expectation would be, presumably, that you would find no solution, as Fermat claims is the case.) After trying various values (without finding a solution), you might be tempted to write a computer program to carry out a more extensive and systematic search for solutions, say trying all values of x, y, z from 1 to 100 (though there are more efficient ways of going about it). But even after several hours of computing you would still have met with no success, and then the futility of such an approach would become all too apparent. For no matter how powerful your computer and how efficient your method, this strategy could *never* succeed in *proving* Fermat's assertion (in this case $n = 4$). For the Last Theorem (with $n = 4$) asserts that *no* triple can be a solution to $x^4 + y^4 = z^4$, which is an assertion about an infinite collection of triples, and no amount of computation will enable you to handle an infinity of cases. Such an approach might succeed in *dis*-proving the Last Theorem, since the discovery of a single solution to the Fermat equation would do that, but it could never prove the theorem. In order to prove the Last Theorem or any one case of it, a more subtle, mathematical approach is required.

As so often in mathematics, the best bet is to look for a proof by contradiction. You want (for $n = 4$) to prove that there is no solution to the equation $x^4 + y^4 = z^4$. So you begin by assuming that there is a solution, say X, Y, Z, and then, on the basis of this assumption, proceed (by means of a mathematical argument) to deduce a contradiction. Once you have obtained your contradiction you will have achieved your goal, since contradictory conclusions can be obtained only from false assumptions – in this case the assumption that a solution did exist.

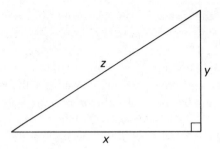

Figure 46. Fermat's result for Pythagorean triangles. By Pythagoras' theorem, the sides of a right-angled triangle satisfy the equation

$$x^2 + y^2 = z^2.$$

The formula for the area of a triangle gives, in this case,

$$\text{Area} = \tfrac{1}{2}xy.$$

Fermat used the method of infinite descent to deduce a contradiction from the assumption that x, y, z were all integers and the area of the triangle was the square of an integer.

The problem facing you is how to deduce a contradiction from your assumption. A particularly powerful method for statements which, like the Last Theorem, involve the natural numbers is the so-called *method of infinite descent*, a method invented by Fermat and (he claimed) used by him as the basis of all his proofs in number theory. An illustration of the method is the very proof Fermat scribbled in the margin of the *Arithmetica*, mentioned a moment ago. It involves what are known as Pythagorean triangles. (You will see presently how it relates to the case $n = 4$ of the Last Theorem.)

For obvious reasons, a triangle is called *Pythagorean* if it is right-angled and all three sides have integer length. (In other words a Pythagorean triangle is one whose sides form a Pythagorean triple.) What Fermat proved is that the area of such a triangle can never be a square (i.e. the square of an integer). His argument runs as follows.

Suppose there were a Pythagorean triangle whose area were a square. Let x, y, z be the lengths of the sides of the triangle, with z the hypotenuse (see Figure 46). Thus, by Pythagoras' theorem, x, y, z satisfy the identity

$$x^2 + y^2 = z^2.$$

Let the area of the triangle be u^2, where u is an integer. Using the formula which says that the area of a triangle is half the product of the base and the height, we find that

$$u^2 = \tfrac{1}{2}xy.$$

By means of a rather ingenious argument, Fermat was then able to derive another set of positive integers $X, Y, Z,$ and U such that

$$X^2 + Y^2 = Z^2, \quad U^2 = \tfrac{1}{2}XY, \quad Z < z.$$

(The argument is given in detail in Chapter 1 of Harold Edwards' book on Fermat's Last Theorem – see the list at the end of this chapter). The sought-after contradiction follows easily from this. The numbers X, Y, Z, U have all the properties that x, y, z, u have, so the same argument may be applied to them to obtain another four positive integers X_1, Y_1, Z_1, U_1 such that

$$X_1^2 + Y_1^2 = Z_1^2, \quad U_1^2 = \tfrac{1}{2}X_1Y_1, \quad Z_1 < Z.$$

Similarly there must exist another four positive integers X_2, Y_2, Z_2, U_2 such that

$$X_2^2 + Y_2^2 = Z_2^2, \quad U_2^2 = \tfrac{1}{2}X_2Y_2, \quad Z_2 < Z_1.$$

And so on, *ad infinitum*. This process is known as *infinite descent* because the positive integers $z, Z, Z_1, Z_2, \ldots$ are getting smaller all the time (i.e. $z > Z > Z_1 > Z_2 > \ldots$). But this is the contradictory situation: there can be no infinite descending sequence of positive integers – eventually you get to 1 and then you must stop. The conclusion is that there cannot be a Pythagorean triangle whose area is the square of an integer.

Though there is no evidence that he actually made the connection, it seems likely that Fermat constructed the above proof in order to prove the case $n = 4$ of the Last Theorem. All that is required in order to deduce that case from the result on Pythagorean triangles is the following simple but ingenious trick.

Suppose that there is a solution to the equation $x^4 + y^4 = z^4$. In terms of this solution, set $a = y^4, b = 2x^2z^2, c = z^4 + x^4,$ and $d = y^2xz$. Then, making repeated use of the familiar algebraic identity

$$(r + s)^2 = r^2 + 2rs + s^2,$$

you get (check this for yourself!)

$$a^2 + b^2 = (z^4 - x^4)^2 + 4x^4z^4$$
$$= z^8 - 2x^4z^4 + x^8 + 4x^4z^4$$
$$= (z^4 + x^4)^2$$
$$= c^2.$$

Also,

$$\tfrac{1}{2}ab = \tfrac{1}{2}y^4 2x^2z^2 = (y^2xz)^2 = d^2.$$

Thus $a^2 + b^2 = c^2$, and $\tfrac{1}{2}ab = d^2$. But this is precisely the situation which the argument above proved impossible. Hence the assumption that the equation $x^4 + y^4 = z^4$ has a solution must have been false. This completes the proof.

It follows immediately that the Last Theorem is true for n equal to any multiple of 4. For if the equation

$$x^{4k} + y^{4k} = z^{4k}$$

had a solution $x = a$, $y = b$, $z = c$, then a^k, b^k, c^k would be a solution of the equation $x^4 + y^4 = z^4$, which has just been proved to be impossible.

More generally, if the Last Theorem can be proved for any given exponent m, then it will be true for all multiples of m. Thus, as every integer greater than 2 is divisible either by a prime greater than 2 or by 4 (or both), then in trying to prove the Last Theorem it is necessary to consider only those cases for which n is a prime greater than 2 (i.e. an *odd* prime) or else $n = 4$. Since the case $n = 4$ has just been disposed of, the problem reduces to the case where n is an odd prime p.

More advanced work on the problem often splits the odd prime case into two subcases. First of all note that, as with $n = 2$ (Pythagorean triples), if x, y, z is a solution to the equation

$$x^n + y^n = z^n,$$

then any multiple of x, y, z will also be a solution, so what is really at issue is whether or not, for a given n, there exists a *primitive solution* – a solution x, y, z having no common factor. For a given odd prime p now, the *first subcase* of the Last Theorem (for p) says that there is no primitive solution to the equation

$$x^p + y^p = z^p$$

for which p divides none of the numbers x, y, z. The *second subcase* says that there is no primitive solution for which p does divide one of x, y, z. Obviously, for a given p, the Last Theorem for that p is equivalent to both subcases holding for that p. The point of splitting up the assertion into two subcases is that it enables some progress to be made on a sort of 'divide and rule' basis. This issue will be returned to later in the chapter.

As mentioned above, there is no evidence that Fermat did prove the Last Theorem for $n = 4$. His proof of the result about Pythagorean triangles outlined above suggests that he probably did – certainly the step from there to the $n = 4$ case is one which was well within his capabilities. At any rate, most sources are content to credit him with the result. A similar cloud of uncertainty also lies over the proof of the next case to be solved, $n = 3$. Though this is almost universally credited to Leonhard Euler, the only published version of his proof likewise contains a 'gap'.

The Case $n = 3$

In a letter to Christian Goldbach dated 4 August 1753, Euler claimed that he had succeeded in proving Fermat's Last Theorem for $n = 3$. But in the letter he gave no proof. Nor indeed did he publish any proof until he included one in his book *A Complete Introduction to Algebra*, published in St Petersburg in 1770. Whether the proof he had in 1753 was correct is not known, but certainly the proof that appeared in 1770 contained a serious flaw. As it turned out, in the case $n = 3$ the flaw can be remedied, but in other cases a similar error proves to be insurmountable. Though Euler's argument is too long to give in detail, it is worth outlining in general terms in order to explain just what the error was and why it was to prove so critical to later attempts to prove other cases of the Last Theorem.

As with Fermat's 'proof' of the case $n = 4$, Euler made use of the method of infinite descent. Starting from the assumption that there was a solution x, y, z to the equation

$$x^3 + y^3 = z^3,$$

he was able to deduce that there was another solution X, Y, Z such that $Z < z$. The crux of the argument is to prove the proposition that if p and q are two numbers having no common factor, and if $p^2 + 3q^2$ is a cube, then there must be numbers a and b such that $p = a^3 - 9ab^2$ and $q = 3a^2b - 3b^2$. This proposition is quite correct, and can be proved by applying techniques found elsewhere in Euler's work. But in the proof of the Last Theorem which he published, Euler chose to employ a novel type of argument involving numbers of the form $a + b\sqrt{-3}$ (where a and b are integers), and this is where the error came in.

We can see why Euler found the numbers $a + b\sqrt{-3}$ useful if we expand the expression $(a + b\sqrt{-3})^3$. It is equal to

$$a^3 + 3a^2b\sqrt{-3} - 9ab^2 - 3b^3\sqrt{-3},$$

which can be rearranged as

$$(a^3 - 9ab^2) + (3a^2b - 3b^3)\sqrt{-3}.$$

So if $p = a^3 - 9ab^2$ and $q = 3a^2b - 3b^2$ (as stated in the conclusion of the proposition to be proved), then

$$p + q\sqrt{-3} = (a + b\sqrt{-3})^3.$$

Now, an assumption of the proposition is that $p^2 + 3q^2$ is a cube. This is equivalent to the statement that $(p + q\sqrt{-3})(p - q\sqrt{-3})$ is a cube, so the entire proposition can be rephrased as follows: if p and q have no common factor, and if $(p + q\sqrt{-3})(p - q\sqrt{-3})$ is a cube (in the system of $\sqrt{-3}$ numbers), then $p + q\sqrt{-3}$ must be a cube (again, in the sense that $p + q\sqrt{-3} = (a + b\sqrt{-3})^3$ for some integers a, b).

To prove this reformulated version of the proposition, Euler reasoned as follows. The numbers $a + b\sqrt{-3}$ (for varying integers a, b) form a number system very much like the integers. (See Chapter 3 for a fairly full explanation of this issue.) Now, if m and n are two given integers having no common factor, and if mn is a cube, then both m and n are cubes. By analogy, Euler argued that the same will be true for the $\sqrt{-3}$ system of numbers. Since, as Euler correctly proved, the assumption that p and q have no common factors implies that the $\sqrt{-3}$ numbers $p + q\sqrt{-3}$ and $p - q\sqrt{-3}$ have no common factors (amongst the $\sqrt{-3}$ numbers), the desired conclusion follows at once.

The problem with this argument – and it is a serious one – is that reasoning by analogy with the integers is not at all valid. Just because the $a + b\sqrt{-3}$ number system resembles the integers in many ways (both systems form an integral domain – see Chapter 3), it does not follow that this system has all the properties of the integers. (It does not.) As to the crucial property for Euler's proof, this holds for the integers by virtue of the fundamental theorem of arithmetic, the unique factorization theorem, which says that every integer is a product of a unique collection of primes (and possibly -1). Now, if you have read Chapter 3 you will be aware that the $\sqrt{-3}$ number system does have this property. So Euler's conclusion is valid. But you will also know from Chapter 3 that $\sqrt{-3}$ is one of only nine integer roots which does lead to the unique factorization property, so it is only by a stroke of luck that Euler's analogy argument did not lead to a false conclusion. Had he been trying instead to prove the case $n = 5$ of the Last Theorem, using the numbers $a + b\sqrt{-5}$, his method would have failed. As will be seen presently, the failure of unique factorization was to prove a rock on which many purported proofs were to founder.

Two More Cases: n = 5 and 7

In 1825, Peter Gustav Lejeune Dirichlet (who had just turned 20) and Adrien-Marie Legendre (who was past 70) proved the Last Theorem for the case $n = 5$. Their method was basically an extension of the one Euler had used for $n = 3$. The analogue of the critical equation

$$p + q\sqrt{-3} = (a + b\sqrt{-3})^3$$

was the identity

$$p + q\sqrt{5} = (a + b\sqrt{5})^5.$$

However, in order to prove that $p + q\sqrt{5}$ is a fifth power (for which argument by analogy with the integers is certainly *not* valid) they had to assume not only that $p^2 - 5q^2$ is a fifth power and p, q have no common factors (as with $n = 3$), but also that just one of p, q is even and that q is divisible by 5. No use is made of (the non-existent) unique factorization.

With the case $n = 5$ disposed of, the approach used by everyone so far began to show signs of strain as the demands made on the algebra became more and more severe. In 1832, having failed to get the method to work for $n = 7$, Dirichlet did succeed in proving the case $n = 14$ (a much weaker result, of course). When, in 1839, Gabriel Lamé finally did prove the case $n = 7$, he had to resort to some very ingenious devices very closely tied to the number 7 itself, and there seemed to be little hope of anyone being able to move on to the next case, $n = 11$, without adopting a radically new approach. It was Lamé himself who, in 1847, proposed just such a course of action.

The Cyclotomic Integers and Lamé's Announcement

Lamé's proposal was to try to prove the full Last Theorem by utilizing a complex nth root of unity: that is, a complex number r for which $r^n = 1$ but $r^k \neq 1$ for any positive integer k less than n. (All of this is for any odd prime n.) For any odd prime n (indeed for any odd number n), the number 1 has $n - 1$ complex nth roots. For instance, for $n = 3$ the two complex cube roots of 1 are

$$-\frac{1}{2} + \frac{\sqrt{3}}{2} i, \quad -\frac{1}{2} - \frac{\sqrt{3}}{2} i.$$

(You can check this by cubing each of these complex numbers.)

The point of introducing such an r is this. The proofs of the cases $n = 3$, 4, 5, 7 which had been found up to that time all depended on some algebraic factorization such as, for $n = 3$,

$$x^3 + y^3 = (x + y)(x^2 - xy + y^2).$$

Lamé realized that the increasing difficulty for larger n was caused by the increasingly large degree of one of the factors in such a factorization. By introducing r, it is possible to factor $x^n + y^n$ completely into n factors each of degree 1.

To obtain the factorization, note that the complex numbers $1, r, r^2, \ldots ,$ r^{n-1} are the roots of the complex equation

$$z^n - 1 = 0,$$

and so

$$z^n - 1 = (z - 1)(z - r)(z - r^2) \ldots (z - r^{n-1}).$$

If you now put $z = -x/y$ and multiply both sides of the equation by y^n, you get (since n is odd)

$$x^n + y^n = (x + y)(x + yr)(x + yr^2) \ldots (x + yr^{n-1}).$$

Each of the complex factors of $x^n + y^n$ in the expression above is a special case of numbers of the general form

$$a^0 + a_1 r + a_2 r^2 + \ldots + a_{n-1}r^{n-1},$$

where $a_0, a_1, \ldots , a_{n-1}$ are integers. Numbers of this type – that is, numbers made up of the integers and the powers of r – are nowadays known as *cyclotomic integers*. (All of this is still relative to some fixed odd prime n.) Like the Gaussian integers or the numbers of the form $a + b\sqrt{-3}$ which appeared above, the cyclotomic integers produce a number system which resembles the ordinary integers to some extent. (They constitute a *ring*; see Chapter 3 for the relevant definitions.)

On 1 March 1847, a highly excited Lamé stood up to address the members of the Paris Academy. He had, he said, at long last succeeded in proving Fermat's Last Theorem. His key idea was to work with (what are nowadays known as) the cyclotomic integers. This enabled him to carry out a proof by infinite descent, much like Euler's argument for the case $n = 3$. (Thus a crucial step in his argument was to show that if the factors $x + y, x + yr, \ldots , x + yr^{n-1}$ of $x^n + y^n$ have no common factors, then the equality of $x^n + y^n$ and z^n would imply that each of the factors $x + y$, $x + yr, \ldots , x + yr^{n-1}$ must be an nth power.)

Having completed the description of his purported proof, Lamé concluded by acknowledging that the idea for using the complex numbers in that way had been suggested to him by his colleague Joseph Liouville a few months before. It was Liouville himself who took the floor when Lamé sat down. Was Lamé really justified, he asked, in concluding that each factor of

$x^n + y^n$ was an nth power if all that he had shown was that no two of these factors had a common divisor? The truth of such a step for the ordinary integers was, he pointed out, dependent on the unique factorization theorem, and he knew of no such result for the cyclotomic integers.

Whether Liouville knew in advance of Euler's previous error on this point is not known. At any rate his remarks struck at the very heart of Lamé's argument, and it was a much saddened and embarrassed Lamé who, after several weeks' valiant effort trying to rescue his proof, finally realized the enormity of his mistake. 'If only you had been in Paris or I had been in Berlin, all of this would not have happened', he wrote to his friend Dirichlet in Berlin. In point of fact, Lamé's acute embarrassment would have been avoided had he only known of some work which had been published by one Ernst Eduard Kummer some three years previously. (Though to be fair to Lamé, for reasons known only to himself, Kummer had chosen the incredibly obscure journal *Gratulationschrift der Universität Breslau zur Jubelfeier der Universität Königsberg* as the resting place for his paper.)

Kummer's Work and Ideal Numbers

In his 1844 paper, Kummer had proved that unique factorization is usually false for the cyclotomic integers, a result which completely destroyed Lamé's purported proof of the Last Theorem. But by 1847, when Lamé and the rest of the mathematical world learnt of these results, Kummer had developed an impressive new theory which showed that there was a way to modify the concept of unique factorization in order that a reasonable 'number theory' could still be obtained for the cyclotomic integers. The basis of his theory was the introduction into the arithmetic of the cyclotomic integers of what he called *ideal prime factors* – a step that is somewhat analogous to the introduction of the imaginary number i into the arithmetic of the ordinary integers. Using Kummer's *ideal numbers*, many of the consequences of unique factorization for the integers may be proved for the cyclotomic integers (and for other number systems such as $a + b\sqrt{-3}$ that arise in proving various cases of Fermat's Last Theorem).

Kummer's work marked the biggest single advance made in the study of Fermat's Last Theorem from its inception until the 1983 result mentioned at the start of the chapter. His 1847 results proved the Last Theorem for all prime exponents less than 37 (thereby establishing the result for *all* exponents less than 37), and for all prime exponents less than 100 with the exception of 37, 59, and 67. All this was just a few years after mathematicians had struggled with proofs for $n = 5$ and $n = 7$.

Moreover, his critical new concept of *ideal numbers* turned out to be an extremely powerful and wide-ranging one which was to lead to a more general concept of an *ideal* and an entire branch of mathematics – *ideal theory* – a theory whose rudiments are nowadays taught as a matter of routine to all university students of mathematics. Indeed, dramatic though the application of Kummer's ideal numbers to Fermat's Last Theorem was, it is the concept of an *ideal* itself that has proved to be the most important aspect of Kummer's work as far as the rest of mathematics is concerned.

In fact, just as the long-term significance of Kummer's work was not its consequence for the Last Theorem, neither did his work stem from an attempt to prove Fermat's claim. Like Gauss (see Chapter 3), Kummer had been working on the problem of higher reciprocity laws to generalize the quadratic reciprocity law proved by Gauss, work which was to lead to his proving in 1859 a powerful general result on the problem. That having been said, it should be noted that there is a close connection between Fermat's Last Theorem and the higher reciprocity laws.

Regular Primes

Kummer's work was particularly valuable as far as the Last Theorem is concerned because it provided an arithmetical condition that an odd prime exponent has to satisfy in order that the Last Theorem be true for that exponent. That is, if an odd prime p satisfies Kummer's condition, then the equation

$$x^p + y^p = z^p$$

has no solution. Nowadays, primes which satisfy Kummer's condition are known as *regular primes*. Of the primes less than 100, only 37, 59, and 67 fail to be regular, as Kummer himself verified in 1847.

What, then, is a regular prime? The concept is closely connected with that of *class number*, described in Chapter 3. A *regular prime* is one which does not divide the class number of the associated cyclotomic number field. Which definition would require considerable explanation were it not for the fact that there is an alternative (though equivalent) definition involving much simpler concepts. Recall from Chapter 3 that e is a standard mathematical constant whose infinite decimal expansion begins $2 \cdot 718\ 28\ \dots$, and that for any number t the value of e^t is given by the infinite sum

$$e^t = 1 + \frac{t}{1!} + \frac{t^2}{2!} + \frac{t^3}{3!} + \dots .$$

The *Bernoulli numbers*, B_k, are defined as the coefficients in the infinite sum

$$\frac{t}{e^t - 1} = 1 + B_1 \frac{t}{1!} + B_2 \frac{t^2}{2!} + B_3 \frac{t^3}{3!} + \dots .$$

The values of the Bernoulli numbers behave quite irregularly. B_k is zero for all odd k except $k = 1$, for which $B_1 = -\frac{1}{2}$. The first few values of B_k for even k are

$$B_2 = \frac{1}{6}, \quad B_4 = -\frac{1}{30}, \quad B_6 = \frac{1}{42}, \quad B_8 = -\frac{1}{30},$$

$$B_{10} = \frac{5}{66}, \quad B_{12} = -\frac{691}{2730}, \quad B_{14} = \frac{7}{6},$$

$$B_{16} = -\frac{3617}{510}.$$

In terms of the Bernoulli numbers, a prime p is *regular* if it does *not* divide the numerators of each of the numbers $B_2, B_4, \dots , B_{p-3}$. (So p will be *irregular* – that is, will fail to be regular – if p does divide the numerator of at least one of the Bernoulli numbers $B_2, B_4, \dots , B_{p-3}$.)

The definition of regularity in terms of Bernoulli numbers provided a way of checking the regularity of a given prime by computation, though using the above definition directly is not very efficient and in practice various facts about the Bernoulli numbers are used to derive more manageable methods. (One problem facing the would-be regularity computationer is that some very large numbers appear as numerators in Bernoulli numbers. For example,

$$B_{34} = \frac{2\,577\,687\,858\,367}{6}$$

is still manageable by hand, but B_{220} has 250 digits!)

The first calculations to determine the regular and the irregular primes were carried out by Kummer himself, as mentioned above. He went as far as 164, finding that the only irregular primes below this were 37, 59, 67, 101, 103, 131, 149, and 157. (In each case Kummer had to show that the prime in question divided the numerator of an appropriate Bernoulli number. For example, 37 divides the numerator of B_{32}, 59 divides the numerator of B_{44}, and 157 divides the numerators of both B_{62} and B_{110}.)

Then, during the 1930s, Stafford and Vandiver used desk calculators (together with some new methods for checking regularity and irregularity) to test all the primes up to 617. In 1954, with the advent of electronic computers, Lehmer and Vandiver took the computation as far as 4001, and various people subsequently helped to reach 30 000. At this point, in 1976, armed with both an IBM 360–65 and an IBM 370 computer, Samuel Wagstaff at the University of Illinois determined the regularity status of every prime below 125 000.

On the basis of all this numerical work it seems that about 60% of all primes are regular. To be precise, for large N the observed ratio is

$$\frac{\text{Number of irregular primes less than } N}{\text{Number of primes less than } N} = 0{\cdot}39.$$

A plausible though non-rigorous argument produced by Siegel in 1964 showed that the above ratio 'ought to be' $1 - 1/\sqrt{e}$. To two places of decimals this works out to be (wait for it) $0{\cdot}39$.

And yet despite the apparent preponderance of regular primes, it is still not known for sure whether there are infinitely many of them. Surprisingly, in view of the numerical evidence, it is known that there are infinitely many irregular primes. This was proved by Jensen in 1915. So the seemingly larger set could turn out to be finite, whilst the apparently smaller set is already known to be infinite!

The Current Position

Kummer's 1847 result showed that if p is a regular odd prime, then the Last Theorem is true for exponent p. But what happens when p is an irregular prime? Here Kummer's result does not help us. (It does not, of course, follow that the Last Theorem is false in this case. All that happens is that the specific argument Kummer used is no longer applicable. The result itself could still be true, for some other reason.) In later years Kummer established for prime numbers more general, though less concise, conditions than regularity, which still implied the Last Theorem. These conditions included the irregular primes 37, 59, and 67, so, quite fittingly, he was able to lay claim to all cases of the Last Theorem up to 100. Since then, even more inclusive conditions (on primes) have been found, so that the computation carried out by Wagstaff in 1976 actually verified the Last Theorem for all primes (and hence all exponents) up to 125 000.

Other information currently available on the Last Theorem concerns the subdivision of each case into two subcases, mentioned earlier. For a given odd prime p, the first subcase says that there is no primitive solution x, y, z to the equation

$$x^p + y^p = z^p$$

for which p divides none of the numbers x, y, z, whilst the second subcase says there is no primitive solution for which p does divide one of the three solution numbers. It is with the first subcase that some significant progress has been made over the years.

In 1832 (well before Lamé and Kummer did their work), the French mathematician Sophie Germain proved that if p is an odd prime such that $2p + 1$ is also prime, then the first subcase of the Last Theorem holds for p. (What this says, then, is that the Fermat equation for p might have a solution, but p would have to divide one of the solution numbers.) Though there are many primes p for which $2p + 1$ is prime (e.g. $p = 3, 5, 11$), and to which Germain's result therefore applies, it is not known whether there are infinitely many.

Legendre subsequently extended Germain's ideas to prove the first subcase for any prime p such that one of

$$4p + 1, \quad 8p + 1, \quad 10p + 1, \quad 14p + 1, \quad 16p + 1$$

is prime. This was sufficient to establish the first subcase for all prime exponents p less than 100, though Kummer's results superseded this of course.

Other results obtained over the years show that the first subcase holds for all primes satisfying various criteria. One of these, obtained by Mirimanoff in 1910, is that p is of the form $2^a 3^b \pm 1$ or of the form $\pm 2^a \pm 3^b$, where a and b are non-negative integers. Since this includes the case of the Mersenne primes (Chapter 1), we know that the first subcase holds for the largest known prime number, the 65 050-digit number $2^{216091} - 1$.

By 1982, Lehmer had shown that the first subcase holds for all primes below 6 billion.

Then in 1985 Adleman, an American, Fouvry, a Frenchman, and Heath-Brown, an Englishman, used a generalization of the Germain criterion to prove for the first time that the first subcase of the Last Theorem holds for infinitely many primes. (Despite all the progress, it is still possible that the full Last Theorem holds only for a finite number of exponents.)

And what then of Faltings' dramatic 1983 result mentioned at the beginning of the chapter? He proved that for each exponent n greater than 2, the Fermat equation

$$x^n + y^n = z^n$$

has at most a *finite* number of primitive solutions. The proof earned him a Fields Medal in 1986.

Whether this will eventually lead to a complete proof of the Last Theorem remains to be seen, but the step it makes from a potential infinitude of solutions to an unknown but finite number is an enormous one. (But notice that use of the word 'unknown' in the last sentence. Faltings' result gives no indication of the maximum number of solutions there might be – only that it is finite.)

In fact the above result is a special case of a more general result which Faltings proved, known as the *Mordell conjecture*. In 1922, Lewis Mordell conjectured that any irreducible polynomial in two variables with rational coefficients, having genus greater than or equal to 2, has at most a finite number of (rational) solutions. (If you have come across the word 'genus' in mathematics before, you will realize that the Mordell conjecture is part of the subject known as topology, discussed in Chapter 10.) Since the polynomial

$$x^n + y^n = 1, \tag{7}$$

with $n \geqslant 3$, satisfies the hypothesis of the Mordell conjecture, it follows at once that this equation has at most a finite number of rational solutions. But since any integer solution to the equation

$$x^n + y^n = z^n \tag{8}$$

will produce a rational solution to Equation (7) (divide both sides of Equation (8) by z^n), with different primitive solutions to Equation (8) giving different rational solutions to Equation (7), this implies at once that Equation (8) has only a finite number of primitive integer solutions.

The Future

All of which leaves us where? Fermat's Last Theorem is known to be true for all exponents up to 125 000, but it is not known to be true for an infinite number of exponents, except in the highly restrictive first subcase. Beyond the limits of what is known there may be one or more prime exponents p (greater than 125 000) for which it is false. For such a p there can be only a finite number of primitive solutions. If p is less than 6 billion, then (since the first subcase holds for such a p) at least one of the numbers in any solution will be divisible by p, so numbers in excess of $125\,000^{125000}$ will be involved in the equation. On the other hand, if p is greater than 6 billion then even more astronomical numbers will arise. So for all practical purposes the Last Theorem is 'true'. But of course, for the mathematician this does not bring the matter to an end at all. The problem of the Last Theorem will not be settled until a rigorous proof or disproof has been obtained. And at the moment it is not clear that any of our present knowledge will be of much use in achieving this goal. It may be that an entirely new approach is required, in which case the Last Theorem may once again lead to significant developments in other fields of mathematics.

What does seem highly likely (though not absolutely certain) is that if a solution is ever found it will involve a great deal more than 'elementary' considerations. This means that the many amateur (like Fermat?) mathematicians who regularly claim to have proved the Last Theorem are almost invariably mistaken. In fact close examination of such proofs usually reveals that their argument does not even settle the case $n = 3$, which Euler proved in 1753. And yet every year a new crop of such 'proofs' appears, giving the Last Theorem the distinction of being the theorem for which the greatest number by far of false 'proofs' have been claimed.

A great many of these 'proofs' find their way to the Mathematics Institute of the University of Göttingen in West Germany. For along with a gold medal and a prize of 3000 francs offered by the French Academy in 1816, the first person to prove Fermat's Last Theorem would win the Wolfskell Prize. When in 1908 this was first offered (by the Royal Academy of Science in Göttingen, in accordance with the will of one Dr Paul Wolfskell), it was worth 100 000 marks. The various, often dramatic changes in the German currency since then have resulted in the current prize being fixed at just over 10 000 deutschmarks. (The prize is due to be withdrawn on 13 September 2007, if it has not been won by then.)

Though there are a number of strictly enforced stipulations concerning submissions for the prize, the Mathematics Institute at Göttingen still receives an average of one solution a week, which it is obliged to evaluate. This is not as bad as during the first year the prize was offered, when 621 entries were sent in!

The chances of success for the untrained amateur (let alone the experienced professional) are, then, extremely slight. But for all that, few mathematicians would wish to discourage anyone from trying. For mathematics is done, above all, for enjoyment, and who would want to deny another his or her personal satisfaction? If you try, and (as is extremely likely) you fail, you can at least console yourself with the knowledge that a great many of the most famous mathematicians in history have also failed to solve this most tantalizing of problems. But if you try and are successful ...

Suggested Further Reading

There are (at least) two excellent, comprehensive sources of information on Fermat's Last Theorem. The book *Fermat's Last Theorem*, by Harold M. Edwards (Springer-Verlag, 1977) begins with three fairly 'elementary' chapters before plunging into the difficult concepts that are so characteristic of most of the work done on the problem this century.

And *13 Lectures on Fermat's Last Theorem*, by Paulo Ribenboim (Springer-Verlag, 1980), although it does not offer the reader quite the same gentle introduction, is again an excellent source of information.

Considerable coverage of Fermat's Last Theorem and related topics is provided in *Algebraic Number Theory*, by I. N. Stewart and D. O. Tall (Chapman and Hall, 1979), a book aimed at the undergraduate student of mathematics.

At a similar level there is also the excellent *Elementary Number Theory*, by David Burton (Allyn and Bacon, 1980), which covers not only Fermat's Last Theorem but also a whole range of other topics in number theory.

9 Hard Problems About Complex Numbers

A Complex Subject

For many readers this will be the most difficult chapter in the book. Not because the mathematics is intrinsically any harder than in other chapters, but because of the degree of abstraction involved. True enough, numbers – both natural and complex – form the core of the subject. But the essential task of *complex analysis* (the name *complex function theory* is used almost synonymously) and the closely allied field of *analytic number theory* (which is the application of the results and techniques of complex analysis to the study of the natural numbers) is to root out and exploit the deep structure and interconnections that lie beneath what – on the evidence of the description of complex numbers given in Chapter 3 – seems a fairly simple notion. To carry out such a study requires some very abstract mathematical techniques which are not familiar to many people outside mathematics itself. And, unfortunately, pictures do not help since the subject is not very visual (unlike topology, the subject-matter of Chapter 10, where equally difficult and unfamiliar ideas can nevertheless be conveyed – at least in simple cases — using pictures and diagrams). But it is an important field, and one in which recent years have seen some significant advances, so it ought not to be ignored. Moreover, if you persevere with this chapter you will discover that out of all the abstraction come some remarkable insights into such familiar concepts as fractions and prime numbers. (It is assumed that you have already read the introductory account of complex numbers given in Chapter 3.)

Though the three problems that form the core of this chapter are all highly abstract, you should not think that complex analysis has no applications outside mathematics. Far from it. Ever since the first work on the subject by Augustin Cauchy in 1825, connections with the outside world have always been there. Riemann's work on the subject, following closely after Cauchy's, showed how complex-function theory could greatly assist in the solution of problems in physics, and further work on the so-called 'integral transforms' (such as the ubiquitous Fourier transform) made the connection even more apparent.

The two-dimensional nature of complex numbers (see Chapter 3 for a discussion of the complex plane) enables them to be used to study problems in two dimensions in much the same way that problems in one dimension can be handled using real numbers. Since real-life problems in three dimensions which have a symmetrical nature (such as the flow of liquid through a circular pipe) often reduce to a mathematical problem in two dimensions, complex analysis is relevant to both the physicist and the engineer.

The Russian mathematician N. Y. Joukowski (1847–1921) used complex analysis to specify the shape of an aerofoil (i.e. the cross-section of an aircraft wing) and to study the flow pattern around it, thereby revolutionizing aircraft design; since then complex-function theory has become of central importance in the description of all forms of fluid flow and in the design of cars and ships. In 1920, scientists at Bell Laboratories in the USA made systematic use of complex-function theory in the design of the filters and high-gain amplifiers which made long-distance telephone calls possible. (If you know an electronics engineer, ask him about the importance of the Nyquist criterion for the stability of feedback amplifiers. This is a direct application of complex analysis.) Complex analysis is, in short, now an indispensable aid; there can be very little of present-day science and technology that is not dependent on complex numbers in one way or another.

So just what is this subject called complex analysis? A good first attempt at an answer would be that it is the extension of the methods of the calculus (differentiation, integration, infinite sums, and so on) from the more familiar setting of the real numbers into the realm of the complex numbers. But like all first attempts, this answer is at the same time helpfully suggestive and misleading, because the whole apparatus of the calculus takes on an entirely different form when developed for the complex numbers. Concepts that with the real numbers appear quite distinct from one another may turn out to be closely related when complex numbers are introduced into the picture.

One instance of this, already mentioned in Chapter 3, is Euler's identity

$$e^{\pi i} = -1.$$

Another (closely related) example is provided by the equation

$$e^{xi} = \cos x + (\sin x)\,i$$

(for any real number x), which relates e, i, and the familiar trigonometric functions sine and cosine. In fact there is nothing to prevent you from applying the sine and cosine functions themselves to complex numbers. Of course, you cannot calculate, say, $\sin(3 + 4i)$ in terms of right-angled triangles as you can with real numbers. Once you decide to work with complex numbers you have to be prepared to go wherever the theory takes you, and for the trigonometric functions this turns out to be into the realm of *infinite series*. Just as the infinite sum

$$e^x = 1 + \frac{x}{1!} + \frac{x^2}{2!} + \frac{x^3}{3!} + \dots$$

gives a valid answer whether x is real or complex, so too do the infinite-series equations

$$\sin x = x - \frac{x^3}{3!} + \frac{x^5}{5!} - \frac{x^7}{7!} + \dots,$$

$$\cos x = 1 - \frac{x^2}{2!} + \frac{x^4}{4!} - \frac{x^6}{6!} + \dots.$$

If x is real, each of these infinite sums will give exactly the same answer you would obtain from the usual, geometric definition (provided that you regard the 'angle' x as being measured not in degrees but in *radians*, where 1 radian equals $180/\pi$ degrees, i.e. about $57 \cdot 3$ degrees). But there is nothing to stop you using these equations when x is complex.

By simple algebraic manipulation of the above three expressions, you can obtain the further formulas

$$\sin x = \frac{1}{2i}(e^{xi} - e^{-xi}), \quad \cos x = \tfrac{1}{2}(e^{xi} + e^{-xi}),$$

where again x may be real or complex.

One of the first tasks of complex analysis is to check that all the above manipulations involving infinite series are allowable. As was made clear in Chapter 2, infinity has to be handled with care, no less so when complex numbers are involved as well.

Integration* turns out to be remarkably different when performed for complex functions. Of course, because complex numbers are two-dimensional you cannot simply integrate from one number a to another number b as with real numbers, as in, for example

$$\int_0^1 x^2 \, dx \;=\; \frac{1}{3}.$$

Instead you have to integrate along a path (in the complex plane). For example, you might want to integrate a function around a circle. What, for instance, would the answer be if you were to integrate the (complex) function $1/(x - a)$ over a circular path C (where a is a complex constant)? (The integral is written as

$$\int_C \frac{1}{x - a} \, dx.$$

Such integrals are sometimes referred to as line integrals.) The result is quite unexpected, especially to those who remember how difficult integration can be for real functions. If the number a corresponds to a point in the complex plane *inside* the circle C, the answer is $2\pi i$; if the number a lies *outside* the circle, the answer is 0. The surprising thing about this, of course, is that the size and position of the circle are not at all relevant, and the only way in which the constant a affects the answer is in whether it lies inside the circle or outside it. (Though this particular example was chosen to provide such a striking result, it is typical of how complex functions take on a life of their own, quite different from what we have come to expect from real numbers.)

More surprises are in store when complex-function theory is applied to the study of the natural numbers, as will become clear as this chapter unfolds. (It is worth noting that the integral mentioned above plays a significant role in this study – though not one that can be explained in the account given here.)

*If you are not familiar with this concept in the calculus for real numbers, you can simply skip the rest of this section and the other occasional references to integration later in the chapter.

Some Number Fun

A fraction h/k is called a *proper fraction* if it lies between 0 and 1 and if h and k have no common factors. For example, 1/2, 3/4, and 7/8 are all proper fractions; 2/4, 3/9, and 3/2 are not. For any number n, the *Farey sequence of order n*, F_n, is the sequence of all proper fractions with denominators which do not exceed n, together with the 'fraction' 1/1, arranged in increasing order. So, for example, F_5 is the sequence

$$\frac{1}{5}, \frac{1}{4}, \frac{1}{3}, \frac{2}{5}, \frac{1}{2}, \frac{3}{5}, \frac{2}{3}, \frac{3}{4}, \frac{4}{5}, \frac{1}{1},$$

and F_7 is the sequence

$$\frac{1}{7}, \frac{1}{6}, \frac{1}{5}, \frac{1}{4}, \frac{2}{7}, \frac{1}{3}, \frac{2}{5}, \frac{3}{7}, \frac{1}{2}, \frac{4}{7}, \frac{3}{5}, \frac{2}{3}, \frac{5}{7}, \frac{3}{4}, \frac{4}{5}, \frac{5}{6}, \frac{6}{7}, \frac{1}{1}.$$

It is not clear exactly who first thought of looking at such sequences. The first to have proved genuine mathematical results about them seems to have been Haros, in 1802. Farey stated one of Haros' results without proof in an article written in 1816, and when Cauchy subsequently saw the article he discovered a proof of the result and ascribed the concept to Farey, thereby giving rise to the name 'Farey sequence'.

The result proved by Haros and then stated by Farey is that if you take any three successive fractions from a Farey sequence, a/d, b/e, c/f, then $b/e = (a + c)/(d + f)$. For example, the 10th, 11th, and 12th members of F_7 are 4/7, 3/5, and 2/3, and

$$\frac{4 + 2}{7 + 3} = \frac{6}{10} = \frac{3}{5}.$$

The other result proved by Haros is that if a/c, b/d are successive members of a Farey sequence, then $bc - ad = 1$. Taking F_7 as an example again, the 6th and 7th members are 1/3 and 2/5, and

$$2 \times 3 - 1 \times 5 = 6 - 5 = 1.$$

It is possible to deduce either of the above two statements from the other, which means that in order to prove them both you need prove only one of them. Both present quite a challenging exercise in algebraic manipulation. (If this is not to your taste, you can at least check the results for a few other Farey sequences, of course.)

For any number n, let $A(n)$ denote the number of terms in the Farey sequence F_n. Thus $A(5) = 10$ and $A(7) = 18$. Suppose you now take the interval from 0 to 1 on the real line and divide it into $A(n)$ equal segments (see Figure 47). The points which divide the interval in this way are the points $1/A(n)$, $2/A(n)$, $3/A(n)$, and so on up to $(A(n) - 1)/A(n)$. Since the terms in the Farey sequence F_n are unequally spaced between 0 and 1, many of the numbers in the sequence will not coincide with the equally spaced points dividing up the interval. Let d_1 be the amount by which the first term of the Farey sequence F_n differs from $1/A(n)$, and d_2 the amount by which the second term differs from $2/A(n)$, and so on, up to $d_{A(n)-1}$. (It does not matter which of each pair is the greater – the difference between them is what counts.) Then let $D(n)$ be the sum of all the numbers $d_1, d_2, \ldots, d_{A(n)-1}$.

To take a simple example, F_4 consists of the numbers

$$\frac{1}{4}, \frac{1}{3}, \frac{1}{2}, \frac{2}{3}, \frac{3}{4}, \frac{1}{1}.$$

Thus $A(4) = 6$. The points which split the interval from 0 to 1 into six equal segments are $1/6$, $2/6$, $3/6$, $4/6$, $5/6$. So d_1 is the difference between $1/4$ and $1/6$, namely $1/4 - 1/6 = 1/12$; d_2 is the difference between $1/3$ and $2/6$, which is 0; d_3 is the difference between $1/2$ and $3/6$, also 0;

Figure 47. The Farey sequence F_4. The members of this sequence are indicated by the arrows, showing their positions relative to the five points which split the interval from 0 to 1 into six equal pieces. The numbers $d_1, d_2, \ldots, d_5$ measure the difference between the Farey fractions and the respective division points. $D(4)$ is the sum of these differences.

likewise $d_4 = 0$; and $d_5 = 5/6 - 3/4 = 1/12$. So,

$$D(4) = \frac{1}{12} + 0 + 0 + 0 + \frac{1}{12} = \frac{1}{6}.$$

(At this stage you should try working out $D(5)$ for yourself.)

In a paper published in 1924, J. Franel and E. Landau investigated the behaviour of the function $D(n)$ as n ranges over all the natural numbers. (They carried out their investigation using algebraic techniques rather than by calculating lots of Farey sequences arithmetically.) In particular, they considered the statement that if r is any real number greater than $\frac{1}{2}$, then there is some constant C such that $D(n)$ is always (i.e. for every n) less than Cn^r. What they proved is that this deceptively simple-looking statement is equivalent to (i.e. is another way of expressing) what is to this day generally acknowledged by professional mathematicians to be the single most important unsolved problem in the field: the *Riemann hypothesis*.

The Most Important Unsolved Problem in Mathematics

To the layman, the most famous unsolved problem in mathematics is undoubtedly Fermat's Last Theorem, considered in Chapter 8. But fame does not always accord with importance. Ask any professional mathematician what is the single most important open problem in the entire field, and you are virtually certain to receive the answer 'the Riemann hypothesis'. Obviously the great British mathematician G. H. Hardy (see Chapter 4) thought so. When faced with a ferry-crossing from Scandinavia to England on a day when the weather conditions on the North Sea were unusually fierce, he sent a postcard to a colleague (no doubt with the origins of Fermat's Last Theorem in mind) bearing the message: 'Have proved the Riemann hypothesis. Yours, G. H. Hardy.' Hardy's reasoning was that God could not possibly allow him to die with the undeserved credit of proving so important a result, and would therefore ensure his safe return home. In the event Hardy (who was, by the way, a devout atheist!) got across safely, and 'Hardy's Last Theorem' did not come into being. The Riemann hypothesis remains unproved to this day.

The story begins around 1740, when Euler introduced into mathematics the so-called *zeta function*, defined for real numbers *s* greater than 1 by the infinite sum

$$\zeta(s) \;=\; \frac{1}{1^s} + \frac{1}{2^s} + \frac{1}{3^s} + \frac{1}{4^s} + \dots$$

(ζ is the Greek letter 'zeta'; the name 'zeta function' has no more significance other than the fact that it is generally denoted by this letter.) For *s* less than or equal to 1, the infinite sum has an infinite answer, so $\zeta(s)$ is not defined for such an *s*. But for any *s* greater than 1 the infinite sum has a definite, finite value. Euler proved that for any such *s* the value $\zeta(s)$ is equal to the infinite product

$$\frac{1}{1 - (1/2)^s} \times \frac{1}{1 - (1/3)^s} \times \frac{1}{1 - (1/5)^s} \times \frac{1}{1 - (1/7)^s}$$

$$\times \frac{1}{1 - (1/11)^s} \times \dots \;,$$

where the (infinite) product is over all numbers of the form

$$\frac{1}{1 - (1/p)^s},$$

where *p* is a prime number. There are two things which make this result an amazing one. Firstly, it shows that the zeta function is closely related to those decidedly basic and concrete entities, the prime numbers. And secondly, the relationship between the two is one which involves the infinite to a very great degree. Obviously there is more to the prime numbers than meets the eye!

For the next episode in the saga, we switch (temporarily) from the zeta function to the distribution of the prime numbers themselves. As you proceed up through the natural numbers, at first the primes appear to be very numerous (for instance, exactly half the first ten numbers beyond 1 are prime), but later on they begin to thin out. And yet there seems to be no overall pattern to the way in which the primes are distributed among the rest of the numbers. For instance, between 9 999 900 and 10 000 000 there are nine prime numbers, but of the next hundred integers, from 10 000 000 to 10 000 100, there are only two: 10 000 019 and

10 000 079. In fact there are runs of integers of all lengths which contain no prime numbers. (For any value of N, the run from $N! + 2$ to $N! + N$ cannot include any primes, as is easily seen.) Another example of the seemingly random way in which the primes appear is that there are many examples of pairs of 'twin primes', primes which differ by 2, such as 3 and 5, 11 and 13, or 10 006 427 and 10 006 429, and these twin-prime pairs seem to occur in a random fashion. (It is conjectured that there are infinitely many such twin-prime pairs, but this has never been proved.)

Nevertheless, there is a form of order underlying the seemingly chaotic manner in which the primes appear. It concerns the behaviour of the function $\pi(n)$, which tells you the number of primes less than or equal to n (see Chapter 1). In his book *Essai sur la Théorie des Nombres* (1798), Legendre observed that $\pi(n)$ is approximately equal to the number

$$\frac{n}{\log_e n - 1 \cdot 083\,66}.$$

(Throughout this chapter, $\log_e n$ will denote the natural, or base e, logarithm of n.) There is, as it turns out, nothing special about the number $1 \cdot 083\,66$ here. Legendre obtained his result by examining tables of primes up to 400 000, and simply chose this number to give as good an answer as possible.

At about the same time as Legendre was working on his book, the fourteen-year-old Gauss also began to investigate the function $\pi(n)$. He observed (though he did not publish the fact until 1863) that $\pi(n)$ is approximated by the number $n/\log_e n$, and also by the number

$$\mathrm{Li}\,(n) \; = \; \int_2^n \frac{1}{\log_e x}\,\mathrm{d}x.$$

(The function Li is the 'logarithmic integral' function. Don't worry if its definition means nothing to you: it is included for 'completeness' only.) Table 2 gives the values of the various approximating functions for values of n up to 100 000 000. From this table it would appear that $\mathrm{Li}(n)$ gives a much better approximation to $\pi(n)$ than do either of the others, and indeed in 1896 it was shown by Charles de la Vallée Poussin that this is so for *all* values of n from a certain point on.

In fact, and this is a slight but interesting digression, from Table 2 it seems that $\mathrm{Li}(n)$ is always slightly larger than $\pi(n)$, and if you were to tabulate further it is virtually certain that this would continue to be what

n	$\pi(n)$	$\dfrac{n}{\log_e n - 1\cdot08366}$	$\dfrac{n}{\log_e n}$	$\mathrm{Li}(n)$
1000	168	172	145	178
10000	1229	1231	1086	1246
100000	9592	9588	8686	9630
1000000	78498	78534	72382	78628
10000000	664579	665138	620420	664918
100000000	5761455	5769341	5428681	5762209

Table 2. The distribution of primes. This is an expansion of Table 1 (in Chapter 1), and shows the number of primes $\pi(n)$ smaller than n for various values of n, along with three classical approximating functions to $\pi(n)$.

you would observe. It would be a sceptical person indeed who did not then conclude that $\mathrm{Li}(n)$ always approximates $\pi(n)$ on the large side. But if you did make such a conclusion you would be wrong! For in 1914 the English mathematician J. E. Littlewood (a colleague of G. H. Hardy's) showed that the difference $\mathrm{Li}(n) - \pi(n)$ changes from positive to negative *infinitely many times* as n runs up through the positive integers, so there will certainly be values of n for which $\mathrm{Li}(n)$ is smaller than $\pi(n)$. Indeed, such an n will, as S. Skewes showed in 1955, have to appear somewhere before the number

$$e^{e^{e^{79}}} \quad \text{(approximately } 10^{10^{10^{34}}}\text{),}$$

a number of incomprehensibly large magnitude and subsequently termed the Skewes number. Considerably smaller, but still well beyond human grasp, is the number $1\cdot65 \times 10^{1165}$ which Lehman showed in 1966 could be substituted in place of the Skewes number (in the sense that $\mathrm{Li}(n) - \pi(n)$ will change sign somewhere below that number). Smaller still is the number $6\cdot69 \times 10^{370}$, and in 1986 H. J. J. te Riele demonstrated that a sign change will occur for some n below this bound. But these are still enormously large numbers, and this is what has led people to conclude that it may never be possible to discover an actual number n for which $\mathrm{Li}(n)$ is less than $\pi(n)$. (Certainly a computer search made as far as one billion (10^9) failed to produce such a number.)

But now we return to the main story. In 1896, the Frenchmen J. Hadamard and C. de la Vallée Poussin independently succeeded in proving conclusively what all the evidence suggested: that as n increases, the value of $n/\log n$ becomes closer and closer to $\pi(n)$ in such a way that no matter how close you want the approximation to be – other than exact equality – you can achieve such accuracy by choosing a large enough n. (It follows that $\mathrm{Li}(n)$ also becomes 'arbitrarily close' to $\pi(n)$ as n gets larger.) This celebrated result, which shows once and for all that there is a definite, mathematical pattern to the way the prime numbers appear, is known as the *prime number theorem*. (Note, however, that the pattern behind the primes involves infinity and the concepts of the calculus.) The work of both mathematicians was dependent on a remarkable eight-page paper written by the German Bernhard Riemann in 1859, the title of which translates as 'On the number of primes less than a given magnitude'. In this, his only paper on number theory, Riemann instigated some lines of research which to this day are still proving extremely fruitful. Indeed, the publication of his paper could be said to mark the beginning of the entire field of analytic number theory, in which the powerful techniques of the calculus are brought to bear on problems about the natural numbers.

The key idea introduced by Riemann in his study of the distribution of the primes was to extend the zeta function $\zeta(s)$ so that, instead of being restricted to the real numbers greater than 1, s is allowed to be any complex number $s = a + bi$ (except $s = 1$). This cannot be done simply by allowing s to be complex in the original Euler definition

$$\zeta(s) \;=\; \frac{1}{1^s} + \frac{1}{2^s} + \frac{1}{3^s} + \frac{1}{4^s} + \;\dots\;.$$

Instead you have to make use of a rather sophisticated technique known as analytic continuation (which will not be described here). For the benefit of those readers who are able to understand the various notions involved (and no doubt for the bafflement of the rest), the formula which Riemann obtained for his extended zeta function in terms of a line integral is

$$\zeta(s) \;=\; \frac{\prod(-s)}{2\pi i} \int_c \frac{(-x)^s}{e^x - 1} \frac{dx}{x},$$

where the integral is taken over the path C which runs left from ∞ along the positive real axis, stops just short of the origin, circles round the origin in an anticlockwise direction, and then goes back along the positive real axis to ∞. The product $\prod(s)$ is given by

$$\prod(s) = \lim_{N \to \infty} \frac{N!}{(s + 1)(s + 2) \dots (s + N)} (N + 1)^s,$$

for all s which are not negative integers.

The extended function is known as the *Riemann zeta function*. It turns out to be of fundamental importance throughout number theory, and entire books have been written about it. (One is H. M. Edwards' 300-page *Riemann's Zeta Function*; as you might gather from the definition of the extended zeta function given above, such books are strictly for the expert.)

For s equal to any one of the numbers -2, -4, -6, and so on, $\zeta(s)$ is equal to 0. Another way of saying this is that the even negative integers are *zeros* of the zeta function. There are infinitely many other complex numbers s for which $\zeta(s) = 0$, all of which have their real part between 0 and 1 (i.e. they are of the form $s = a + bi$, where a is between 0 and 1). The *Riemann hypothesis* is a conjecture Riemann made in his paper about these complex zeros of the zeta function. He suggested (on hardly any real evidence, as far as anyone can see) that all complex zeros of the zeta function have their real part exactly equal to $\frac{1}{2}$ (i.e. if $\zeta(s) = 0$, then $s = \frac{1}{2} + bi$ for some number b).

Why is this hypothesis so important? Well, as Riemann demonstrated in his paper, there is a very close relationship between the zeros of the zeta function and the properties of the function $\pi(n)$. It was this connection which led Hadamard and de la Vallée Poussin to their respective proofs of the prime number theorem. (Their proofs depended on the *connection*, which is known to be true, and not on Riemann's hypothesis about the zeros, whose truth remains unknown to this day.) The connection also lies behind a great deal of other known facts about primes, including the work on Skewes' number mentioned earlier. If the Riemann hypothesis does turn out to be true and the zeros of the zeta function really are so well ordered, then the connection with the function $\pi(n)$ will enable even more information about the prime numbers to be deduced than is at present known, and this is what makes it such an important problem for the mathematician at large. So just what is known about the likelihood of the hypothesis being true?

Well, before describing the considerable work that has been done on the Riemann hypothesis since it was first put forward, it is perhaps worth while quoting just one specific approximation to $\pi(n)$ which involves the zeta function explicitly. (This one requires only the original Euler form of the zeta function.) It is (and again many readers may be unfamiliar with the notation, in which case don't worry, just read on):

$$R(n) = 1 + \sum_{k=1}^{\infty} \frac{1}{k\zeta(k+1)} \frac{(\log n)^k}{k!}.$$

(The sum in this equation is an infinite sum – from $k = 1$ to $k = \infty$.)

Table 3 shows just what an astonishingly good approximation to $\pi(n)$ is the quantity $R(n)$.

n	$\pi(n)$	$R(n)$
100000000	5761455	5761552
200000000	11078937	11079090
300000000	16252325	16252355
400000000	21336326	21336185
500000000	26355867	26355517
600000000	31324703	31324622
700000000	36252931	36252719
800000000	41146179	41146248
900000000	46009215	46009949
1000000000	50847534	50847455

Table 3. The distribution of primes. This table begins where Table 2 ends. For these larger values of *n* the function *R(n)*, which is defined using the Riemann zeta function, provides an extremely good approximation to the prime-distribution function *π(n)*.

The Riemann Hypothesis

What then is the evidence to support Riemann's conjecture that if $\zeta(s) = 0$ for a complex number s, then s is necessarily of the form $\frac{1}{2} + bi$? As mentioned earlier, it is certainly the case that any complex zero will have its real part between 0 and 1, and $\frac{1}{2}$ lies midway between these two bounds, but Riemann surely had greater evidence than that. But whatever the grounds for his assertion, no one alive today knows what they were. As will soon become clear, he was not in a position to sit down and calculate many zeros, and, even if he had, the example concerning the sign of the expression $\mathrm{Li}(n) - \pi(n)$ shows just how unreliable numerical evidence can be in mathematics. G. H. Hardy was able (and it was not at all easy!) to prove that infinitely many zeros have their real part equal to $\frac{1}{2}$, but this does not preclude there being infinitely many more which do not. Apart from that there really is little more to go on other than the numerical evidence accumulated as a result of some prodigious calculations over the years. (Though such evidence can never prove the hypothesis, if it turns out to be false it is always possible – but unlikely – that numerical computation will reveal a number which demonstrates this falsity. For all it would take to disprove the Riemann hypothesis would be the discovery of just one zero whose real part was not equal to $\frac{1}{2}$. This provides one justification for taking a computational approach. Another reason is that it provides an excellent testing ground for new numerical algorithms which may turn out to have other uses.)

There are a number of techniques for computing the values of b for which the complex number $\frac{1}{2} + bi$ is a zero of the zeta function. There are also ways of calculating the total *number* of zeros (as opposed to calculating the zeros themselves) whose imaginary part lies within any specified range. By combining two such calculations it is then possible to check the Riemann hypothesis for any given finite range. The first person to adopt a computational approach to the Riemann hypothesis was J.-P. Gram. In 1903, using a standard technique called Euler–Maclaurin summation to calculate the zeros, Gram found the first fifteen values of b for which $\zeta(\frac{1}{2} + bi) = 0$. The first ten he computed to six decimal places, the first one being $b = 14 \cdot 134\,725$ and the tenth $b = 49 \cdot 773\,832$. The remaining five he computed to only one decimal place, the eleventh being $b = 52 \cdot 8$ and the fifteenth $b = 65 \cdot 0$. Knowing that any complex zero must have its real part between 0 and 1, he then went on to show that there are exactly ten zeros whose imaginary part lies between 0 and 50. Since his earlier list of zeros of the form $\frac{1}{2} + bi$ has exactly ten entries whose imaginary part lies within the range 0 to 50, it follows that this list comprises all the zeros with an imaginary part in that range. In other words, his computations proved that the Riemann hypothesis is true in the range 0 to 50. (The 'range' in this context always refers to the size of the imaginary part of the zero.)

Using similar (though somewhat improved) techniques, R. Backlund in 1918 verified the hypothesis for all zeros in the range 0 to 200, and (using still further improvements to the method) in 1925 J. J. Hutchinson raised the upper limit to 300. In 1936 Titchmarsh and Comrie used an improved method devised by Carl Siegel to compute 1041 zeros; all were of the form hypothesized by Riemann.

After the Second World War, electronic computers were brought to bear on the problem. During the 1950s, using the Siegel method for computing zeros of the form $\frac{1}{2} + bi$ together with a new method suggested by Alan Turing for determining the number of zeros in a given range, D. H. Lehmer began to take the search much further, and by 1966 R. S. Lehman had taken the number of computed zeros to 250 000. Within a few years of that, J. B. Rosser and his colleagues took the total to 3 500 000. By 1983, after work by J. van de Lune and H. J. J. te Riele, the entire region from 0 to 119 590 809·282 had been investigated and all the (exactly) 300 000 001 zeros in this region had been found to be of the type Riemann predicted. In 1985 the same pair took the calculation as far as the first one-and-a-half billion zeros, again without finding one which countered the Riemann hypothesis.

So all of the numerical evidence supports Riemann's hypothesis. If it is false, it must fail for numbers way beyond the range generally considered by anyone other than the professional pure mathematician, which puts the problem on a par with Fermat's Last Theorem (see Chapter 8). Despite all the work and all the evidence, no one really knows whether it is true or not. But as if to caution once again against the temptation to base conclusions on numerical evidence, within the past few years a conjecture closely related to the Riemann hypothesis *was* resolved. And in this case the numerical evidence turned out to be entirely misleading. The fate of the *Mertens conjecture* should serve as a warning to all who 'assume' that the Riemann hypothesis has to be true.

The Mertens Conjecture

If you take any natural number n, then, by the fundamental theorem of arithmetic, either n is prime or else it can be expressed as a product of a unique collection of primes. For instance, for the first five non-primes,

$$4 = 2 \times 2, \quad 6 = 2 \times 3, \quad 8 = 2 \times 2 \times 2,$$
$$9 = 3 \times 3, \quad 10 = 2 \times 5.$$

Of these, 4, 8, and 9 have prime decompositions in which at least one prime occurs more than once, while in the decompositions of 6 and 10 each prime occurs once only. Numbers divisible by the square of a prime (such as 4, 8, 9) are called *square-divisible*. Numbers not so divisible are called *square-free*. (Thus in the prime decomposition of a square-free number, no prime will occur more than once.)

If n is a square-free natural number that is not prime, then it is a product of either an even number of primes or an odd number of primes. For example, $6 = 2 \times 3$ is a product of an even number of primes, while $42 = 2 \times 3 \times 7$ is a product of an odd number of primes. In 1832 A. F. Möbius introduced the following simple function (denoted by the Greek letter 'mu' and called the *Möbius function*) to indicate what type of prime factorization a number n has. For $n = 1$, let $\mu(n) = 1$, a special case. For all other n, $\mu(n)$ is defined as follows:

if n is square-divisible, then $\mu(n) = 0$;
if n is square-free and the product of an even number of primes, then $\mu(n) = 1$;
if n is either prime, or square-free and the product of an odd number of primes, then $\mu(n) = -1$.

So, for example, $\mu(4) = 0$, $\mu(5) = -1$, $\mu(6) = 1$, $\mu(42) = -1$, and you may work out further values for yourself.

Now, for any number n let $M(n)$ denote the result of adding together all values of $\mu(k)$ for k less than or equal to n. For example:

$$M(1) = \mu(1) = 1,$$

$$M(2) = \mu(1) + \mu(2) = 1 + (-1) = 0,$$

$$M(3) = \mu(1) + \mu(2) + \mu(3)$$
$$= 1 + (-1) + (-1) = -1,$$

$$M(4) = \mu(1) + \mu(2) + \mu(3) + \mu(4)$$
$$= 1 + (-1) + (-1) + 0 = -1,$$

$$M(5) = 1 + (-1) + (-1) + 0 + (-1) = -2,$$

and you can check for yourself that

$$M(6) = -1, \quad M(7) = -2, \quad M(8) = -2,$$
$$M(9) = -2, \quad M(10) = -1, \quad M(11) = -2,$$
$$M(12) = -2, \quad M(13) = -3, \quad M(14) = -2,$$
$$M(15) = -1, \quad M(16) = -1, \quad M(17) = -2,$$
$$M(18) = -2, \quad M(19) = -3, \quad M(20) = -3.$$

(Question: What is the first value of n for which $M(n)$ is zero again? Or positive again?)

All of which seems a pleasant enough bit of fun, but hardly likely to be related to the most important unsolved problem in mathematics, you might think. (Though having read the section on Farey sequences, you may not be quite so sure.) Well, as will become clear in a moment, the behaviour of the function $M(n)$ turns out to be very closely related to the location of the zeros of the Riemann zeta function.

The connection was certainly known to T. J. Stieltjes. In 1885, in a letter to his colleague C. Hermite, he claimed to have proved that no matter how large n may be, the size of $M(n)$ (i.e. neglecting any minus sign) is always less than $\sqrt{n}$:

$$|M(n)| < \sqrt{n} \tag{9}$$

(The two vertical bars here are the standard notation for indicating the suppression of any minus sign in the expression within. Thus $|-10| = 10$, $|5| = 5$, and so on.) If what Stieltjes claimed had been true, the truth of the Riemann hypothesis would have followed at once. In fact, the Riemann hypothesis follows from the existence of any constant A such that the inequality

$$|M(n)| < A\sqrt{n}$$

holds for all n. Needless to say in view of what we have seen in the previous section, Stieltjes was wrong in his claim, though at the time this was not at all clear. (For instance, when Hadamard wrote his now-classic and greatly acclaimed paper proving the prime number theorem in 1896, he mentioned that he understood Stieltjes had already obtained the same result using the inequality (9), and excused his own publication on the grounds that Stieltjes' proof had not yet appeared!) The fact that Stieltjes never did publish a proof might well suggest that he eventually realized his error. At any rate, in 1897, F. Mertens produced a 50-page table of values of $\mu(n)$ and $M(n)$ for n up to 10 000, on the basis of which he was led to conclude that inequality (9) was indeed 'very probable'. As a result this conjecture is nowadays known as the Mertens conjecture.

In a series of papers spread over the period 1897 to 1913, von Sterneck published additional values of $M(n)$ for selected values of n up to 5 million, and found that they too all satisfied the Mertens conjecture. Indeed beyond $n = 200$ they all satisfied the stronger inequality

$$|M(n)| < \tfrac{1}{2}\sqrt{n},$$

but his conjecture that this was always the case was proved to be false by Jurkat in 1960. The smallest value of n for which $|M(n)| \geqslant \tfrac{1}{2}\sqrt{n}$ is $n = 7\,725\,038\,629$, which gives the value $M(n) = 43\,947$.

Then, in 1979, Cohen and Dress computed $M(n)$ for all n up to 7·8 billion, and observed that all their values satisfied the inequality

$$|M(n)| < 0\cdot 6\sqrt{n},$$

which again seemed to suggest that the Mertens conjecture might be true. (To be fair, it should be mentioned that a great deal of other evidence suggested otherwise!) But in fact it is not true, and in October 1983 Hermann te Riele and Andrew Odlyzko brought eight years of collaborative work to a successful conclusion by proving just that.

Their result was obtained by a combination of classical mathematical techniques and high-powered computing. Moreover, their collaboration was typical of an increasing trend in present-day research in that for most of the time each remained at his own home base (te Riele at the Mathematical Centre in Amsterdam, and Odlyzko at Bell Laboratories, New Jersey) and communicated with the other via electronic mail.

So how did they set about obtaining their result? Certainly not by finding a number n for which $|M(n)| \geqslant \sqrt{n}$. To date no such number has been found, and the available evidence suggests that there is no such n below 10^{30}. Rather, the key to their proof was provided by a result obtained by A. E. Ingham in 1942.

To begin with, note that the inequality $|M(n)| < \sqrt{n}$ can be rewritten as

$$\frac{|M(n)|}{\sqrt{n}} < 1.$$

What Ingham did was to show how to define a certain function $h(x)$ which has the following property: for any (real) number x, no number less than $h(x)$ can be greater than every value of $|M(n)|/\sqrt{n}$. So in order to disprove the Mertens conjecture, what you need to do is find a number x such that $h(x) > 1$. Suppose, for example, you can find an x with $h(x) = 1 \cdot 06$. Then the Ingham result tells you that no number less than $1 \cdot 06$ is greater than every value of $|M(n)|/\sqrt{n}$. In particular, 1 (which is less than $1 \cdot 06$) is not greater than every value of $|M(n)|/\sqrt{n}$. This contradicts the Mertens conjecture. (But notice that it does not provide you with a value of n for which the Mertens inequality fails.)

So, in order to dispose of the Mertens conjecture, te Riele and Odlyzko set out to find a value of x for which $h(x)$ is greater than 1. This turned out to be a rather difficult task.

The definition of the function $h(x)$ involves some very abstract mathematics, including the Riemann zeta function. In particular, computation of $h(x)$ for a given value of x involves the fairly accurate calculation of a considerable number of zeros of the zeta function. (Moreover, Ingham's result for the function $h(x)$ depends upon the fact that all the zeros calculated are in accordance with the Riemann hypothesis – though if you expect this hypothesis to be true you are not likely to worry about this side of things, and in any case the hypothesis has been checked way beyond the range required to disprove the Mertens conjecture.) These computations are certainly far too time-consuming to be performed by anything but the most powerful computers available. And even then, when you have written a program which will perform the calculation of $h(x)$ for any given x, you are faced with the task of finding an x for which the answer is greater than 1. This part is even harder. The function $h(x)$ exhibits a frustrating tendency to produce values which are almost always well below 1. Indeed, by 1979, the best that te Riele had been able to achieve was a value of 0·86. Finding a suitable x was, it seemed, like looking for a needle in a haystack. At the time, te Riele concluded that the problem was outside the range of existing computers.

The breakthrough which led to the solution of the problem turned out to come not in the form of new computer technology, but from a powerful (and, it transpired, widely applicable) new algorithm discovered by Lenstra, Lenstra, and Lovász in 1981. When applied to the Mertens conjecture, this new technique (which will not be described here) provided just the additional computational facility that was required in order to find an appropriate value of x. Now te Riele and Odlyzko had a method for searching through their haystack.

The first part of the proof was to calculate the zeros of the zeta function that were likely to be required in order to compute the values of the $h(x)$ function they would encounter in their search. A CDC CYBER 750 computer at the Amsterdam Mathematical Centre was run for 40 hours in order to do this, obtaining 2000 zeros accurate to 100 decimal digits. Then, using the powerful new algorithm mentioned above, a CRAY-1 computer at Bell Laboratories was run for 10 hours until a value of x was found for which $h(x)$ was greater than 1. In fact, for the value of x found, $h(x) = 1·061\ 545$. The Mertens conjecture had finally been laid to rest. For the record, the value of x that did the trick was the following giant, which has 65 digits before the decimal point:

$$- 14\,045\ 289\ 680\ 592\ 998\ 046\ 790\ 361\ 630\ 399\ 781\ 127$$
$$400\ 591\ 999\ 789\ 738\ 039\ 965\ 960\ 762·521\ 505$$

Of course, not only does this result disprove the Mertens conjecture, it shows that the inequality

$$|M(n)| < 1·06\sqrt{n}$$

is not always valid. But what of the inequality

$$|M(n)| < A\sqrt{n}$$

for other values of the constant A? If for any value of A such an inequality were valid (for all n), then the Riemann hypothesis would follow. In order to prove that an inequality of this form is false using the method adopted by te Riele and Odlyzko, you would have to find a value of x for which $h(x)$ is greater than A. Both te Riele and Odlyzko believe that this is theoretically possible, however large A may be – which is to say that they believe the function $h(x)$ achieves arbitrarily large values. But, they caution, the best that could be hoped for in practice, given present-day algorithms and computer technology, is to find a value of x for which $h(x)$ is around $1\cdot5$. Even getting up to 2 is, they conclude, beyond our present capabilities.

So with the Mertens conjecture proved false, where does that leave the Riemann hypothesis? Exactly where it was. Had the Mertens conjecture been true (as Stieltjes had once thought), the truth of the Riemann conjecture would have followed at once. But knowing that the Mertens conjecture is false says nothing about the Riemann hypothesis one way or the other. The two conjectures, Mertens' and Riemann's, are not equivalent.

There is, however, a weakened version of the Mertens conjecture that is equivalent to the Riemann hypothesis: namely that for any real number r greater than $\frac{1}{2}$, there is a constant A such that the inequality

$$|M(n)| < An^r$$

is valid for all n. (In other words, take any power of n greater than the $\frac{1}{2}$ of the Mertens conjecture, and you get an inequality of the appropriate type.) The falsity of this proposition would, of course, imply the falsity of the Riemann hypothesis (just as its truth would imply the truth of the Riemann hypothesis). But that is an altogether different problem from the Mertens conjecture.

The Bieberbach Conjecture

Mathematics, as it is generally presented to the world, is a body of cold, impersonal knowledge. Mathematical truth, uniquely, does not vary from minute to minute, from place to place, or from person to person. But mathematics *as it is developed* is a human endeavour, and therefore subject to a whole range of influences. Though at the end of the day the absolute nature of mathematical truth cannot be denied, it can sometimes take some time to arrive at that day's end.

Imagine for a moment that you are one of the world's leading mathematicians. You have been working hard on a particular problem for many years, and on several occasions you have come close to solving it. One day you receive a 385-page typed manuscript which purports to solve your problem. You check the name at the top, Yes, a respectable mathematician. This is not one of the dozens of 'crank' papers you routinely receive and immediately consign to the waste bin. But this person is 52 years of age, and the conventional wisdom has it (on some pretty convincing statistical evidence) that no mathematician produces much of note beyond the age of 40 or so. Moreover, this particular mathematician has in the past claimed to have solved other problems and in many cases serious errors have been found in the arguments. And now he is claiming to have solved a problem that in its 70-year history has defeated not only yourself – an acknowledged world expert on this particular problem – but also many other leading mathematicians around the world (several of whom have at one time or another mistakenly thought for a short while that they had found a solution). A quick glance through the manuscript reveals that your correspondent is using a complicated method that you and practically everyone else familiar with the problem would regard as most unlikely to be successful.

Faced with such a situation – and no doubt having many other things that you want and need to be getting on with – what would you do? This is the position in which the American mathematician Carl FitzGerald found himself in the spring of 1984. Had his countryman Louis de Branges really done as he was claiming, and managed to solve the Bieberbach conjecture, finding success where so many others had failed? To FitzGerald it seemed unlikely. ('I was not hoping for the proof to be correct,' he wrote later. 'Two of my best papers showed the Bieberbach conjecture was at least close to being true ... I did not want these results superseded.') Similar scepticism was shown by each of the other dozen or so mathematicians across the country to whom de Branges had also sent a copy of his manuscript. So no one read it.

But, as chance would have it, de Branges had a prior arrangement to visit the USSR as part of an exchange programme with the USA, and whilst he was there (during April to June of that year) he was scheduled to lecture to the mathematicians at the University of Leningrad. It was the ideal place to describe his claimed proof. Amongst the mathematicians in his audience would be I. M. Milin, E. G. Emel'anov, and G. V. Kuz'mina, three acknowledged world experts on the problem. Indeed, what de Branges was in fact claiming was to have proved a conjecture proposed in 1971 by Milin himself, and shown by him to imply the Bieberbach conjecture as a consequence. Though they were also sceptical, the Russian mathematicians proved to be a patient audience, sitting through a series of five 4-hour lectures on the purported proof. Their expectation was that an error would be discovered at any moment.

But that moment never came; there was no error. The proof was correct. The next step was to see if it could be simplified, to reduce it from its hitherto daunting length. It could. After some work, the group managed to find some significant modifications to the argument which brought the length down to a mere 13 pages. When these 13 pages were sent out, the rest of the world sat up and took notice – and believed. What de Branges had done, most mathematicians dream of but few ever achieve. He had solved a long-standing open problem that everyone had agreed was 'hard'.

In fact it is the difficulty of the problem that was the main reason for its fame. As far as is known, the Bieberbach conjecture does not lead to a host of other significant developments, as does (say) the Riemann hypothesis. But it is one of those nice 'tidying-up' results that demonstrate the orderly nature of mathematics in general and complex analysis in particular. What it says is described below.

Suppose you have an infinite series of the form

$$B = x + a_2 x^2 + a_3 x^3 + \dots ,$$

where x is a complex variable and the coefficients $a_2, a_3, \dots$ are all complex numbers. (There is no mention of an a_1 since the coefficient of x is 1, i.e. $a_1 = 1$.) For a given value of the variable x, it is possible for the infinite series to produce a finite answer (or 'sum'). In order for this to happen, however, the successive terms in the series will have to grow smaller at a very fast rate – so fast in fact that the effect of there being an infinite number of terms is counteracted.

For example, the series for $\sin x$ given earlier is of the form of B:

$$\sin x = x - \frac{x^3}{3!} + \frac{x^5}{5!} - \frac{x^7}{7!} + \dots$$

(so $a_2 = 0$, $a_3 = 1/3!$, $a_4 = 0$, $a_5 = 1/5!$, $a_6 = 0$, and so on). This infinite series gives a finite answer for any value of x, because the coefficients a_n grow smaller at such a fast rate. (The same is true of the series for e^x, though this is not of the form of B since it begins with a 1. But if you subtract this initial 1 to obtain the series for $e^x - 1$, then you do get a series of the form of B which, like $\sin x$, gives a finite sum for any value of x.)

If, however, the coefficients in the series do not grow small in such a rapid fashion, the existence of a finite sum may well depend on the size of the number you take as the value of x. Small values of x may lead to a finite answer, larger values to an infinite answer (or no answer at all in some cases). In particular, if x is less than 1, then as n increases the size of x^n decreases, and consequently there is a chance of obtaining a finite answer. It all depends on the size of the coefficients a_n. Consequently, when mathematicians study series of the form of B, they frequently restrict their attention to the behaviour of the series for values of x for which x^n grows smaller as n increases; that is, they consider only those numbers x of absolute value less than 1. (The absolute value of a complex number x is its distance from the origin in the complex plane, denoted by $|x|$. This notation has already been introduced for real numbers x, when it signified the suppression of any negative sign. If you think about it for a moment you will realize that the two uses are not contradictory. The new use for complex numbers simply extends the old one. By a straightforward application of Pythagoras' theorem for right-angled triangles, if $x = a + bi$, then $|x| = \sqrt{a^2 + b^2}$.)

Thought of in terms of the complex plane, the set of all those complex numbers x for which $|x| < 1$ will form a circular disc of radius 1 with its centre at the origin, usually referred to simply as *the unit disc*. (So to say that x lies in the unit disc is just a geometric way of saying that $|x| < 1$.)

Suppose now that you have a series of the form of B and that this series gives a finite (complex) answer (which I shall denote by $f(x)$) for every value of x in the unit disc. Thus the series determines a *function* $f(x)$ which to each x in the unit disc assigns a *value* $f(x)$. Now it may be that you can find two different values of x which both give the same answer when you apply the function f. (For instance, suppose that the B series has $a_2 = 1$ and all other coefficients a_n are zero, so that $f(x) = x + x^2$. Then, as you may check for yourself, $f(-1/3)$ and $f(-2/3)$ are both equal to $-2/9$.) If this does not occur (that is, if different values of x always give different answers), then the function $f(x)$ is said to be *one–one* (from the fact that any one value of $f(x)$ comes from only one value of x). Mathematicians give this particular property a name because it is a very useful one. (A bit like the requirement of monogamous marriage in most modern societies.)

Assuming from now on that our B series (whatever it actually is) does yield a one–one function $f(x)$ which has finite values for every number x in the unit disc, it is possible to picture the action of $f(x)$ geometrically in terms of the complex plane. What $f(x)$ does is associate with every point x in the unit disc another point $f(x)$ somewhere in the complex plane. The set of all the values $f(x)$ obtained in this way will be some region (subset) of the complex plane. It is natural to ask how big this region will be. Since the values of x that you start with have to come from the unit disc, and since the unit disc is small compared with the entire complex plane (which stretches out to infinity in every direction), you might be tempted to think that the region of the values $f(x)$ can also be only a small part of the plane. But remember, both the unit disc and the entire complex plane contain infinitely many points, and, as was amply demonstrated in Chapter 2, infinite sets do not at all behave in the same way as do the more familiar finite sets. In this case they most certainly do not. The answer to the above question is that the region of values of $f(x)$ may be *practically the entire complex plane*! Moreover, you don't need a particularly fancy function $f(x)$ to achieve this. The so-called *Koebe function* does it. This can be calculated either from the formula

$$K(x) = \frac{x}{(1 - x)^2}$$

or else from the B series

$$K(x) = x + 2x^2 + 3x^3 + 4x^4 + 5x^5 + \dots .$$

(Since the Koebe function does come from an infinite series of the form of B, and since it is one–one, it is precisely the kind of function we have been looking at.) The set of all values $f(x)$ that the Koebe function gives for x in the unit disc consists of every complex number except for those lying on the real axis to the left of $-\frac{1}{4}$. So, geometrically, this function is pretty well the 'biggest' one possible (where 'big' refers to the size of the region of values).

Since the Koebe function is a sort of 'record holder', it is natural to ask yourself if this property is reflected in the individual coefficients in its B series in any way. For instance, is each coefficient the largest one possible? That is to say: suppose you take any series of the form of B which determines a (finite-valued) one–one function $f(x)$ for all numbers x in the unit disc, does it follow that

$$|a_2| \leqslant 2, \quad |a_3| \leqslant 3, \quad |a_4| \leqslant 4, \dots ?$$

(Or, to put the question the other way round, if your B series has one or more coefficients which *fail* to satisfy the relevant inequality, say if $|a_{163}| = 163 \cdot 5$, then does it follow that either f fails to be one–one or else, even worse, that there is some number x in the unit disc for which the B series fails to give a finite answer?)

This is the question that the German mathematician Ludwig Bieberbach asked himself early this century, and in a paper published in 1916 he conjectured that the answer is 'Yes'. But the only coefficient for which he was able to prove his conjecture was the first one; that is, he did manage to prove that $|a_2| \leqslant 2$.

And that was where the saga began. The conjecture seemed a reasonable enough one, just the kind of 'tidy' result that complex analysis so often yields, and it was sufficiently easy to state to guarantee that there would be no shortage of mathematicians prepared to attempt a proof. But it was to be a long time before that proof was discovered.

After Bieberbach himself, the first result on the conjecture was obtained by another German, Charles Löwner, who in 1923 developed a method which enabled him to prove the conjecture for the coefficient a_3 (i.e. he showed that $|a_3| \leqslant 3$). It was not until 1955 that Garabedian and Schiffer, working in the USA, succeeded in proving the conjecture for the next coefficient, a_4. Then, in 1968, Pederson and Ozawa skipped over a_5 to prove the conjecture for a_6, and in 1972 Pederson and Schiffer teamed up to dispose of the bypassed a_5. In the same year, Ozawa and Kubota proved that $|a_8| \leqslant 8$, missing out a_7, and that was to mark the end of this step-by-step approach.

Slow progress indeed. And moreover there is no possibility that the conjecture could be proved in its entirety by this process of examining each coefficient one by one (since the conjecture is for every one of the infinitely many coefficients). But there was always a chance that such an approach might lead to just the insight required to solve the complex problem. In the event this was to turn out to be true: when Louis de Branges finally proved the Bieberbach conjecture in 1984, his method was indeed based upon some of the early work. Which early work? That of Löwner in 1923, the first to work on the problem after Bieberbach himself! But we are getting ahead of the main story, for whilst slow progress was being made in attacking the coefficients one by one, another approach was also under way.

The idea was to investigate inequalities of the form

$$|a_n| \leqslant Cn,$$

where C is some constant, and to see how small C could be made whilst the inequality was provably valid for all values of n. In order to prove the Bieberbach conjecture, it would be necessary to show that $C = 1$ was possible, but, in the meantime, how close could you get to this value?

The first result of this kind was found by J. E. Littlewood, who in 1925 proved that $C = e$ was possible. (Recall that e is approximately equal to 2·718.) After that various workers steadily lowered this value. In 1956, Milin (mentioned earlier) showed that you could have $C = 1·243$. In 1972, FitzGerald (the recipient of the 385-page manuscript mentioned at the beginning of this section) lowered this to $C = 1·081$, and in 1978 his student David Horowitz took it down to 1·066. They were getting tantalizingly close, but not close enough. And when the complete solution to the Bieberbach conjecture finally came in 1984, it was (as mentioned above) based, not upon this 'reduce C' approach, but upon Löwner's original 1923 work.

In order to prove the conjecture for a_3, Löwner introduced a partial differential equation whose solutions approximate any function which is one–one on the unit disc. The trick now was to translate the mathematics into physical terms, such as the flow of water along a river. The differential equation represents an expanding flow, and it is possible to 'push information' along the flow – in particular estimates of the size of the coefficients in the B series for the function. For his proof, de Branges introduced some auxiliary functions $t_1, t_2, \ldots$ to hold the desired information, each t_k being defined by a differential equation involving all previous functions up to t_{k-1}. The proof of the Bieberbach conjecture (in fact of the stronger Milin conjecture alluded to earlier) then reduced to a proof that these t functions satisfy certain conditions. It was, all in all, a decidedly 'old-fashioned' approach to the problem, in many ways resembling a huge juggling trick as de Branges struggled to keep all the balls in the air. But it worked.

Suggested Further Reading

An introductory account of complex numbers can be found in many elementary books on mathematics, for example *Sets, Functions and Logic*, by Keith Devlin (Chapman & Hall, 1981). Likewise there are many sources of more advanced information about complex analysis, such as *A First Course on Complex Functions*, by G. J. O. Jameson (Chapman & Hall, 1970).

Introduction to Analytic Number Theory, by K. Chandrasekharan (Springer-Verlag, 1968) does exactly what its title says – though the subject-matter is by its very nature not accessible to the non-mathematician.

The classic book on Riemann's zeta function is *Riemann's Zeta Function*, by H. M. Edwards (Academic Press, 1974).

The solution to the Mertens conjecture is described in the research article 'Disproof of the Mertens conjecture', by A. M. Odlyzko and H. J. J. te Riele, in the *Journal für die reine und angewandte Mathematik*, Volume 357 (1985), pp. 138–60.

For a brief account of the Bieberbach conjecture and its solution see the article 'The last 100 days of the Bieberbach conjecture', by O. M. Fomenko and G. V. Kuz'mina, in the journal *The Mathematical Intelligencer*, Volume 8 (1986), pp. 40–7. Another account can be found in Carl FitzGerald's article 'The Bieberbach conjecture: Retrospective', in the *Notices of the American Mathematical Society*, Volume 32 (1985) pp. 2–6.

10 Knots and Other Topological Matters

Boy Scouts, Physicists, and Another Book

How can you tell a reef knot from a granny knot? The average Boy Scout will have no difficulty in answering, but can the mathematician make the distinction? 1984 saw some significant developments affecting this question.

Do physicists use the right kind of mathematics to study the four-dimensional space–time universe we live in? Until 1982 every mathematician would have said, 'Yes, of course, it is the *only* kind.' But now we know better. There are other, quite different kinds of mathematics which apply to a four-dimensional universe – but *only* to a four-dimensional universe: no such special treatment is required for two dimensions, or for three, or for five, six, and beyond. Four-dimensional space is special not only because our universe appears to be four-dimensional, it is mathematically special as well – but in a totally unexpected way.

These are just two of the developments that have taken place in recent years in the vast area of mathematics known as *topology*, a subject now so big that it would be possible to write an entire book entitled 'Topology: The New Golden Age' instead of just this one chapter. Topology, or at least certain aspects of it, now pervades most areas of present-day mathematics, to say nothing of many deep connections with modern physics. The subject has already made one explicit appearance in this book, in the discussion of the four-colour theorem in Chapter 7. (Connections with topics in other chapters were not brought out, but are there none the less.) And yet it is barely a century old. Though some of the ideas go back to Euler and Gauss, it was really only with the work of Henri Poincaré and others in the latter years of the nineteenth century that topology really got under way.

This chapter will deal with just two aspects of topology: *knot theory*, a fascinating though highly specialized part of the subject, and *manifold theory*, the study of the properties of geometric surfaces and generalizations of them. Most topologists would regard manifold theory as the central thrust of their discipline. (One theme that is ignored, though it satisfies all the criteria for inclusion in this book, is catastrophe theory, a form of applied manifold theory developed in the mid 1960s by the topologists René Thom and Christopher Zeeman. My excuse is that a good layman's account of this subject is easily available elsewhere.*)

Though topology is an extremely difficult subject to pursue properly, a facility to visualize geometric objects is all that is required for the interested outsider (yourself, presumably) to grasp the general principles. For this reason the account that follows is highly geometric in its approach.

What is Topology?

Topology is a part of mathematics which is at the same time easy to describe and yet very difficult to appreciate properly. The simple description is that topology is the study of those properties of geometric objects which remain unchanged under continuous transformations. At an intuitive level, a *continuous transformation* (known also as a *topological transformation*) can be thought of as subjecting the object to bending, stretching, compressing, or twisting – or any combination of these, it being assumed that the object being deformed (or *transformed*) is ideally elastic and capable of any degree of such manipulation. The proviso is that points in the object which are close together before the transformation is applied remain close together in the transformed object. When properly formulated – and this is by no means easy – this requirement does not, as it might at first seem, conflict with the object being stretched to any degree whatever. What it does is to prevent any cutting, tearing, or 'gluing' of the object. At least in general terms it does: what *is* allowed is the cutting of the object to allow for some manipulation to be performed that would not otherwise be

Catastrophe Theory, by Alexander Woodcock and Monte Davis (Penguin, 1980).

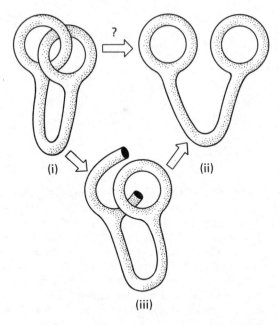

Figure 48. A ring puzzle. Imagine the object in (i) made from a perfectly elastic material. Can you deform it so as to unlink the two rings, as in (ii)? The most obvious way is to make a cut in one of the rings, as in (iii), separate the two rings, and then rejoin the two cut ends. Provided you join the two free ends in exactly the same way as before they were cut (i.e. without twisting one of the two ends round at all), this procedure will be a legitimate topological transformation of (i) into (ii). But in fact it is possible to transform (i) into (ii) without any cutting, simply by manipulating the object in the appropriate manner. Can you see how to do it? A solution is shown in Figure 62.

possible, followed by the gluing together of the cut edges so that points that started out close together are once again close together. For instance, in order to change the object illustrated in Figure 48(i) into that shown in Figure 48(ii) by continuous transformation, it is permissible to cut one of the two interlocked rings and separate it from the other ring, as in Figure 48(iii), and then glue the two cut edges back together again. As the two objects are thus transformable one into the other, they are said to be *topologically equivalent*. (In fact, for this particular example cutting is not necessary. It is possible to transform Figure 48(i) into Figure 48(ii) by means of stretching and bending alone. Can you see how to do it? The solution is shown in Figure 62, at the end of this chapter. What this

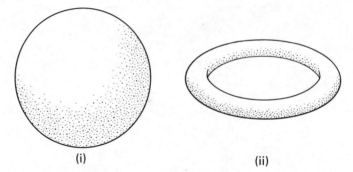

(i)   (ii)

Figure 49. The sphere (i) and the torus (ii). In topology these are both thought of as (hollow) surfaces, not solid objects. Most people would agree that it is not possible to deform either one of these figures into the other by topological means – that the two surfaces are therefore topologically inequivalent. To prove this you would have to show that there is a topological feature of one surface not shared by the other. The 'hole' in the torus would seem to be adequate. The problem is that the hole is a property not of the torus surface, but of the surrounding space. The topological feature which does distinguish between these two figures is the so-called Euler characteristic. Though a genuine property of the surface itself, the Euler characteristic of the torus faithfully corresponds to the hole in (or rather 'not in') the torus.

example shows is that even 'simple stretching' is quite a tricky concept to grasp.)

One of the reasons why the 'stretching–cutting–gluing' description of topology is inadequate is that it really applies only to the study of surfaces (i.e. two-dimensional objects), whereas topology deals with objects with any number of dimensions – three, four, five, and beyond. But even for surfaces there are difficulties. Accustomed as we are to thinking in terms of two-dimensional objects situated in three-dimensional space, it is very easy to confuse the properties of the object with the properties of the space surrounding it. Indeed, the very concept of dimensionality itself often causes confusion. For example, the sphere is regarded in this chapter as a spherical, two-dimensional surface, not as a solid ball. A point on this surface can move in only two independent directions *on the surface*. However, the physical construction of a sphere can be achieved only in a space having at least three dimensions.

To take another example: the sphere and the torus* shown in Figure 49

---

*Like the sphere, this too is to be thought of as a hollow surface.

would seem to be topologically distinct (and indeed they are). By no amount of stretching, bending, or cutting-and-gluing would it seem possible to transform one into the other. (Remember that the only allowable cutting-and-gluing procedure requires the gluing together of all points separated by the cut, so you cannot transform a torus into a sphere by cutting across it to form a cylinder and then gluing together the two ends.) The fundamental difference between the two surfaces would seem to be that the torus has a 'hole' in it, would it not? Well, no, not quite. The torus is a smooth surface having no holes at all. If you were constrained to live on a large torus, you could wander all over its surface without ever encountering a hole. The hole is connected with the way this particular surface is situated in three-dimensional space. Or, to put it another way, the hole is something in the surrounding space, not in the surface. This is not to say that the hole is irrelevant in the topology of the torus, only that it is not a property of the surface itself. There is a topological property that the torus possesses which is closely related to the existence of the hole in the surrounding space, and in a moment we shall see what it is. (It serves to distinguish between the torus and the sphere.)

This distinction between a property of a surface and a property of the surrounding space is a subtle one. The point may become a little clearer when contrasted with the topological notion of an *edge*. Neither the sphere nor the torus has an edge: they are both *closed* (i.e. edge-free) surfaces. A disc has one edge. A disc with a hole in it (such as a gramophone record) has two edges. If you take a strip of paper and glue the ends together to form a cylindrical band (see Figure 50(i)) you get a surface with two edges. If you give the strip a single half-twist before gluing the ends together (Figure 50(ii)), you obtain a surface with just one edge (make one and see!) known as a *Möbius band*, named after A. F. Möbius, whom we met in Chapter 9.

If you were to take two Möbius bands and glue them together edge-to-edge, the resulting figure would be a Klein bottle (see Figure 44 in Chapter 7). However, working in three-dimensional space this construction is not possible, and it can be depicted only by allowing the surface to pass through itself. Incidentally, statements of the form 'the Möbius band (or Klein bottle) is a surface having just one side' (as in Chapter 7) are also a little misleading when used to try to convey the fundamental ideas of the topology of surfaces, since 'sidedness' too depends on regarding the object as sitting inside a space of three or more dimensions (from which lofty standpoint it is possible to look down upon the sides of the surface). Discovering that it is impossible to paint what appears to be the two sides of a Möbius band or a Klein bottle using two different colours (try it – at least for

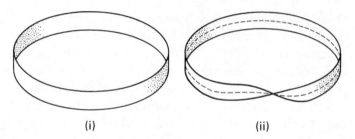

(i)   (ii)

Figure 50.   The cylindrical band (i) and the Möbius band (ii). The cylindrical band is an example of a two-edged, orientable surface; the Möbius band is a single-edged, non-orientable surface. To construct a Möbius band from a strip of paper, simply give the paper a single half-twist before gluing the two ends together. The Möbius band has some curious properties. As an example, you could investigate what happens when the Möbius band is cut 'in half' along the dashed line shown above. The result is not at all what you might expect. You could also make a similar cut on another Möbius band, but this time one-third of the way in from the edge.

a Möbius band – and see what happens) certainly helps you to appreciate the unusual nature of these surfaces, but it does not bring out the genuine topological property of the surface, which in this case is not sidedness, but 'orientability'. So from now on try not to think of surfaces as having 'sides'.

A surface is said to be *orientable* if the notions of 'clockwise' and 'anti-clockwise' are distinguishable in the following sense. Suppose you draw a small circle on the surface (better still, say *in* the surface – no 'sides', remember) and give it a direction, as in Figure 51. Then no matter how you move the circle over (better: *within*) the surface, you will never be able to reverse the direction of the circle. ('Small' as applied to the circle here means small enough to be freely movable all over the surface without getting caught up around any holes, protrusions, or other features.) The cylindrical band (Figure 50(i)) is orientable; the Möbius band (Figure 50(ii)) is *non-orientable*. To see this for yourself, you should construct your bands from some kind of transparent material (the 'acetate sheets' sold for use with overhead projectors are excellent for this purpose), so that when you draw your directed circles they can be seen from both 'sides' and may therefore be regarded as 'in' the surface. Now, starting with a small arrowed circle drawn somewhere on the band, 'move' it around the band by making successive copies (including the arrow) at regular intervals. With the Möbius band (see Figure 52), when you arrive back at your starting place (you might mistakenly think you are on the opposite 'side' from where you began) you will discover that your circle has the arrow pointing the other way round: your sequence of circles show how 'clockwise' and 'anti-clockwise' can be interchanged *without ever leaving the surface*. This is non-orientability. With the cylindrical band, no such reversal is possible; this is orientability.

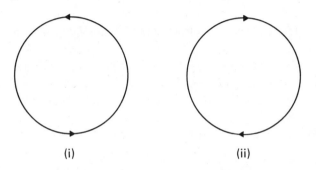

(i)   (ii)

Figure 51.   Orientability. A surface is said to be orientable if it is impossible to turn an anticlockwise-arrowed circle (i) into a clockwise-arrowed circle (ii) by transporting the circle over the surface. Orient-ability is the topological property of surfaces which corresponds to the intuitive notion of 'two-sidedness' (as in the cylindrical band); non-orientability corresponds to 'one-sidedness' (as in the Möbius band).

Figure 52.   Non-orientability of the Möbius band. Construct a Möbius band from a strip of transparent material and follow what happens when a directed circle is moved around the band.

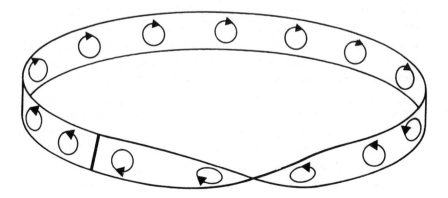

How Do You Do Topology?

A facetious, though highly pertinent reply to this question would be, 'With great care.' The very general kinds of transformation allowed in topology mean that most of the familiar properties of geometric figures no longer apply. In classical geometry (which may be described as the study of properties of objects which remain unchanged under *rigid transformations* – translation, rotation, and reflection) use is made of such concepts as straightness, circularity, angle, length, area, and perpendicularity. None of these makes any sense in topology, as all can be destroyed or altered by means of continuous transformations.

In classical geometry, to verify that two objects are the same you see if it is (theoretically) possible to move one of them by a rigid transformation so as to occupy exactly the same position as the other. This is called *transforming* one object into the other. If such a transformation can be found, then the two objects are said to be 'the same' or 'identical' or, more formally, *geometrically equivalent.* In order to show that two objects are not the same, you would find some specific geometric property which is different for the two, such as one being bigger than the other, or a certain angle being different. All of which is, of course, so familiar as to appear too trivial to mention. After all, we compare objects in this way all the time. But exactly the same procedure is adopted in topology, the only difference being that the transformations and the distinguishing concepts used are different (and not at all as familiar to us).

To show that two objects are *topologically equivalent** you look for a topological (i.e. continuous) transformation which will take one into the other. For instance, a triangle and a circle are topologically equivalent, there being a perfectly obvious way of deforming one to the other. To demonstrate that two objects are topologically distinct (i.e. non-equivalent) you look for some topological property which one has but the other does not: having an *edge*, for example. The disc has an edge, the sphere does not, and so the two are topologically different. (You cannot continuously deform a disc into a sphere or vice versa.) In order to distinguish (topologically) between two closed surfaces (i.e. surfaces having no edge) such as a sphere

*The words 'same' and 'identical' are used by mathematicians, but for the beginner they carry too strong a geometrical connotation and so are avoided here wherever possible.

and a torus, some other topological property has to be found. The 'hole' of the torus will not do, since – as we have seen – this in itself is not a topological property of the surface. Nor will orientability do it, since both surfaces are orientable. (Since the Klein bottle is non-orientable, this concept will however serve to distinguish both the sphere and the torus from the Klein bottle.)

One of the first tasks of topology, therefore, is to find enough topological properties to enable us to distinguish any two non-equivalent objects by showing that one of them has a particular property and the other does not. Any such property must, of course, be the same for all objects which are topologically equivalent to the one considered. That is, a topological property must not be changed by a continuous transformation. (If it were, it would not be a topological property!) For this reason such properties are often called *topological invariants*, a name which is somewhat better than 'property' since many invariants are numerical ones. Orientability is one topological invariant; so too is the number of edges of a surface. (A surface with one edge cannot be equivalent to one with three edges.) There is one further topological invariant which, together with the two just mentioned, suffices to distinguish between all non-equivalent surfaces. But before we look at it, a remark is called for concerning the use of phrases such as '*the* sphere' or '*the* Klein bottle'. Since any two spheres (say) are topologically equivalent (they can differ only in position and size, neither of which is topologically invariant), from the topologist's point of view they are 'identical', and hence it makes sense to speak of '*the* sphere', even if the object you are actually looking at looks like a long sausage. Of course, if two spheres are being considered, such usage of the definite article would be inappropriate, but when the intention is simply to convey the *idea* of a sphere, the phrase 'the sphere' makes sense.

And so to that third invariant of surfaces. In fact you have met it already – it was introduced and used in Chapter 7. As shown in Chapter 7, for any surface, the number $V - E + F$ obtained from the number of vertices (V), edges (E), and faces (F) in a map or network covering the (entire) surface is independent of the actual network chosen and its location on the surface. Moreover, it can be proved (and it is certainly easy to believe) that this quantity remains unaltered by any continuous transformation of the surface, and so is a topological invariant of the surface. It is called the *Euler characteristic* of the surface.

For the sphere the Euler characteristic is 2 (i.e. $V - E + F = 2$ for any network covering the surface of any object topologically equivalent to a sphere). For the torus it is 0, and hence the torus and the sphere are topologically distinct. Incidentally, it is the Euler characteristic that is related to the 'hole' of the torus. The double-torus, for which there are two holes, has Euler characteristic -2, the treble-torus has Euler characteristic -4, and so on; a general formula is given later. For the Klein bottle the Euler characteristic is, as for the sphere, 0, so this invariant does not distinguish between these two objects (but, as has already been mentioned, orientability does the job in this case). The three invariants – number of edges, orientability, and Euler characteristic – are adequate to distinguish all (two-dimensional) surfaces.

Knot Topology

Some topologists might say that knot topology is not 'topology'. And in a sense they are right. Knot theory is a kind of topology, but a very special kind. In knot theory the objects of interest (*knots*) are of necessity objects in three-dimensional space. In two dimensions it would be impossible to construct a knot – there is not enough 'room' to wrap the string, or whatever you are using, around itself. In four or more dimensions there is too much 'room' – any knot would immediately 'fall out' to leave an unknotted string. Moreover, the kind of topological deformations that may be performed on knots specifically exclude the cutting and gluing that is permitted in a general topological transformation. (The reason for this restriction is obvious.)

So what is knot theory? Precisely what its name suggests: a mathematical study of knots. Two simple examples of knots are shown in Figure 53(i) and (ii): the overhand knot and the figure-of-eight. If you take a length of string and form either of these two knots you will discover that they really do knot the string; they do not simply 'tangle' it. The distinction is that a knot can be undone only by a process which involves pulling a free end through a loop at some stage, whereas a tangle can be removed even if the two ends of the string are held fixed. Moreover, the overhand knot and the figure-of-eight seem to be quite different: one cannot be transformed into the other except by threading or unthreading a free end. (If both ends are fixed, no amount of manipulation will turn an overhand knot into a figure-of-eight knot.)

But despite the apparent simplicity of things so far, we are already treading on thin ice. True enough, playing with a piece of string for half an hour

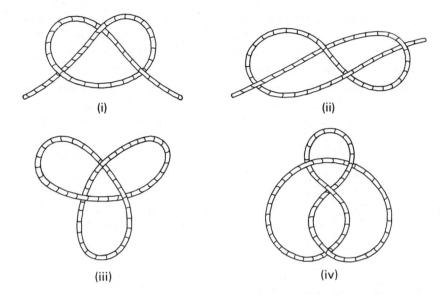

Figure 53. Basic knots: (i) overhand, (ii) figure-of-eight, (iii) trefoil, (iv) four-knot. The first two are types of knot that you might construct using a length of string. However, for a mathematical study the free ends have to be joined together to form a closed loop. Joining the ends of (i) gives (iii); joining the ends of (ii) gives (iv). Drawings (iii) and (iv) are examples of what are known as knot diagrams.

or so might fail to unknot the string or to convert one knot into another. But that does not really prove that the string is knotted or that two knots are really different. It could be that you have simply failed to find the correct sequence of moves. (Remember the puzzle in Figure 48?) And what about very complicated-looking configurations? When a long piece of string or wire gets into a hopeless-looking tangle (as happens with frustrating regularity to the cables on electric lawn-mowers), it is impossible to tell if it is merely tangled (in which case it can be straightened out by carefully pulling the ends) or genuinely knotted. The complicated-looking jumble of wire might well conceal no knot at all, or nothing more than a simple overhand knot. And of course, stage conjurors make extensive use of impressive-looking 'knots' which do not actually 'knot' the rope at all. A potentially very confusing state of affairs indeed, and one which requires a decidedly careful approach – a systematic, mathematical approach. This is what knot theory sets out to do.

Since the object of knot theory is to investigate 'knottedness', the first thing to be done is to get rid of those free ends. Though free ends are essential when you actually want to *tie* a knot, their presence would prevent a mathematical study from ever getting off the ground, since both tying and untying are (cutting-free) topological processes. In other words, from a topological point of view, no string with free ends can be knotted. The simplest solution is to join the free ends together to create a closed (and presumably knotted) loop, and this is what is done in knot theory. When the free ends of the two knots in Figure 53(i) and (ii) are joined, the resulting (mathematical) knots are the *trefoil* and *four-knot*, respectively, shown in Figure 53(iii) and (iv).

A *knot*, then, is just a closed loop (of string, rope, or whatever). (As a special case this definition will include the simple, 'unknotted' closed loop. Known as the trivial knot, this is somewhat analogous to the number 0 in arithmetic or the empty set in set theory.) Two knots are said to be *equivalent* if it is possible to transform one into the other by a topological process that does not involve cutting. The principal aim of knot theory is to find a collection of *knot invariants* which is adequate for a distinction to be made between any two non-equivalent knots (in much the same way as orient-ability and Euler characteristic serve to distinguish between any two non-equivalent closed surfaces). In particular, it should be possible to determine whether a given knot really is knotted or whether it is equivalent to the trivial knot (i.e. a simple loop of string).

The first such investigations seem to have been made by the ubiquitous Gauss. Certainly Gauss's student Listing devoted a large part of his mono-graph *Vorstudien zur Topologie* (1847) to the topic, following which there was quite a lot of work done on the subject. (Though it was not until 1910 or thereabouts that Dehn managed to prove that there is such a thing as a non-trivial knot. Until then it was still a theoretical possibility that no genuine knots exist to be studied!)

Much of the early work, carried out by Kirkman, Tait, Little, and others in the period 1870 to 1900, concerned the classification of the so-called *prime knots* (which, as you might imagine, are knots that cannot be split up into two simpler knots). These were classified according to their *crossing number*. To obtain the crossing number of a knot, first you lay it out flat on a level surface, either physically, if your knot is made of string, or else by means of a 'projection' if it is a conceptual knot. You then manipulate it so that it has as few crossings as possible, and at no single point do three or more lengths of the knot cross over each other. (A crossing is just a point where the string crosses over itself.) Then the total number of points where the string crosses over itself is the *crossing number* of the knot, which is an invariant of the knot. For example, the knots shown in Figure 53(iii) and (iv) are both 'laid out' according to the above procedure, and so by inspection you can see that they have crossing numbers of 3 and 4, respectively. (These knots are also prime.)

As it is a knot invariant, the crossing number may be used to distinguish between non-equivalent knots: if two knots have different crossing numbers, then they cannot be equivalent. The concept is, however, not as useful as might be imagined. For one thing, many different knots can have the same crossing number, so the invariant often fails to detect non-equivalence. And for another thing, it can be quite hard to calculate the crossing number since you have to display the knot in such a way that there are no superfluous loops or crossings, and for even quite simple knots it is usually not at all obvious when this has been achieved!

Nevertheless, by the end of the nineteenth century a great many prime knots with crossing numbers up to 10 had been distinguished and tabulated. What was not known was whether the tables included all knots of each of the crossing numbers, or whether there were any redundancies in the table in the sense that apparently different but in fact equivalent knots were included as separate entries. In 1927, J. W. Alexander* and G. B. Briggs examined this question of redundancy, and managed to prove that there were no duplications in the tables up to and including crossing number 8. Their methods were almost as successful for knots with crossing number 9 as well – only three pairs could not be distinguished by means of the properties that they were using. Subsequent work by Reidemeister took care of these outstanding three pairs. Distinguishing all the tabulated knots with crossing number 10 was accomplished only as recently as 1974, by Perko. As for the question of finding new knots not included in the old nineteenth-century tables, hardly any progress was made until 1960 when John Horton Conway of Cambridge University invented a new and more efficient notation for knots. This enabled him not only to discover some prime knots that the earlier workers had missed, but also to extend the tables to cover prime knots with crossing number 11. There are, it would appear, some 801 prime knots with crossing number at most 11, consisting of one each with crossing numbers 3 and 4, two with crossing number 5, three with 6 crossings, seven with 7, twenty-one with 8, forty-nine with 9, one hundred and sixty-five with 10, and five hundred and fifty-two with

---

*This is not the same Alexander who is said to have 'undone' the famous Gordian knot. That was Alexander the Great, in 333 BC, and he used the method of slicing through it with a sword, which is forbidden to present-day knot theorists.

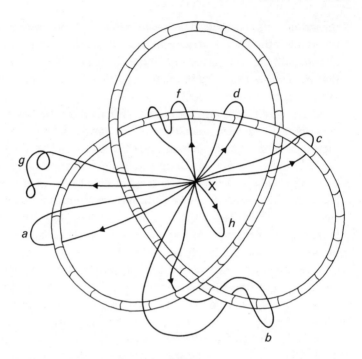

Figure 54. Construction of the knot group for the trefoil knot (see the text for details).

crossing number 11.* (It has not been proved conclusively that the new tables are all-inclusive, though it is believed that they are. Nor has the question of possible redundancies been ruled out. Hence these figures are still not guaranteed to be correct.)

By far the most significant advance in the mathematical theory of knots was the realization that with every knot can be associated a certain group – the so-called *knot group* (for that knot). (See Chapter 5 for an introduction to the mathematical concept of a 'group'.) This is one of those marvellous occasions in mathematics when the concepts and results from one part of the subject are discovered to be useful in another, in this case the application of group theory to knot theory.

---

*The prime knots with crossing numbers 3 and 4 are the two knots shown in Figure 53(iii) and (iv). For pictures of prime knots with up to 9 crossings, see the paper 'On types of knotted curves', by J. W. Alexander and G. B. Briggs, in the journal *Annals of Mathematics*, Volume 28 (1927), pp. 562–86. For knots with 10 crossings, see 'On the classification of knots', by K. A. Perko, in the journal *Proceedings of the American Mathematical Society*, Volume 45 (1974), pp. 262–6.

The idea behind the construction of the knot group is a simple one, and can be explained by using the trefoil as an example (see Figure 54). Start by choosing some point X not on the knot. (Exactly where X is is not important. The final group you get will be the same wherever it lies.) In order to define a group now, we have to do three things: (a) define the objects which make up the group, (b) establish how two objects in the group are combined together to produce a third object in the group, and (c) check that the various group axioms (see p. 112) are satisfied.

The objects that will form the group (for the trefoil as our example) are directed (i.e. arrowed) paths which start at the point X, then wind around the knot to various degrees, and finish at the point X. (The paths may not pass through the material of the knot.) But not all such paths are included in the group. Paths having superfluous loops (i.e. loops which can be untwisted or cut out without affecting the way the path winds around the knot) are omitted. Thus path *g* in Figure 54 will not be in the group, whereas path *a* will. Also, when there are two paths which can be deformed one to the other (without 'passing through' the material of the knot), such as paths *a* and *b*, only one of them is included in the group. But note that paths *c* and *d* are not such a pair, since they pass around the knot in opposite directions, and direction is an important factor. One special path which is also included in the group is the trivial 'path', of zero length. Any path which does not wind around the knot at all, such as path *h*, is, of course, deformable into this trivial path, and hence not included in the group. And that is it. Note that this definition will produce a group which contains infinitely many objects, since there will be paths which wind around a given length of the trefoil any finite number of times before returning to X, and they will all be quite distinct (i.e. not deformable into each other).

Having established what the objects of the group are (and the fact that these may not be quite the kind of objects which you usually think of as forming a group shows just how wide-ranging a concept the 'group' is), the next step is to decide how two such objects are combined, or 'multiplied together'. Given two (directed) paths *p* and *q*, take the path which consists of *p* followed by *q*. It will always be possible to deform this combined path to a path that has been put into the group, because all paths are put into the group except those deformable to one already there. It is this path which will be the 'product' $p * q$ of the two objects *p* and *q* in the group. For example, in Figure 54, $d * d = f$, as is easily seen. In combining two paths, opposite directions cancel out to leave no path at all. For example, paths *c* and *d* will cancel out, a fact which may be written as $c * d = e$, or $d * c = e$, if we use *e* to denote the trivial path.

What has to be done now is to check that these definitions give a system which satisfies the group axioms. It is left to you to do this part. The identity element of the group is, fairly obviously, the trivial path e. Incidentally, for the trefoil at least, the knot group is not commutative, and you should be able to convince yourself of this fact too.

The knot group is (and this at least seems obvious in view of the definition, allowing as it did for deformations of all the paths) an invariant of the knot. Two equivalent knots will give rise to the 'same' group; that is, their knot groups will be exact copies of each other as far as their group behaviour is concerned, though the precise paths which are put into the two groups will very likely be different. (For finite groups, one group will be an 'exact copy' of another if their tables – see Chapter 5 – are the same. For infinite groups, as here, the mathematical notion of an *isomorphism* is required in order to make this precise.) Hence the knot group may, like the crossing number, be used to distinguish between non-equivalent knots. The knot group is, however, a much more powerful invariant. Indeed, the knot group manages to capture so much of the 'knotty structure' of the knot – as seems clear from its definition – that it is only rarely that two different knots turn out to have the same knot group. (But the familiar reef and granny knots are such a pair, so there are no grounds for complacency!) Hence the knot group promises to provide an excellent means of telling knots apart. But there is a problem: how do you obtain a sufficiently simple algebraic description of the group? (The knot group is an infinite, abstract mathematical structure, remember.) Using the description outlined above to define the knot group is obviously not the way to do it – that is even more complicated than the knot itself! Fortunately this problem turns out to have a solution: in 1910, Dehn discovered a way of obtaining from a diagram of any knot a simple and concise algebraic description of the associated knot group.

This means that the knot group does indeed provide an excellent, practical means of classifying knots. It allows the very powerful techniques of group theory to be brought to bear on the problem, sometimes leading to the development of other useful invariants (one of which, the Alexander polynomial, will be described later). In the meantime, one simple direct application of the knot group can be given. Consider the following question: is it possible to untie a given knot (i.e. make it equivalent to a circle) by introducing an additional knot in the string and then unravelling the result? The answer is 'No'. Tying an additional knot in the string will create a bigger and more complicated knot group than the original one, whereas the knot group of the trivial knot (i.e. the circle) is the same as the group formed by the integers with the operation of addition. Since the knot groups are different, the knots cannot be equivalent.

Another way of looking at knots was discovered in 1935 by Seifert. He devised a way to construct, for any given knot, an orientable (i.e. 'two-sided') surface which has the knot as its only edge. Now, a standard result in the topology of surfaces (see the next section) is that any one-edged, orientable surface is topologically equivalent to a disc with a certain number of 'handles'. A 'handle' is added to a disc by cutting two holes in the disc and stretching a cylindrical tube from one to the other – see Figure 59. The number of handles on the disc is called the *genus* of the original surface; like the Euler characteristic, genus is a topological invariant of the surface. Using Seifert's construction, it is therefore possible to associate with any knot a certain natural number, namely the genus of the Seifert surface for that knot. This number is called the *genus of the knot*. (Actually there is a slight complication, in that Seifert's method can give different surfaces – with different genuses – starting from the same knot. So what you do is take the smallest genus number that arises from the knot in this way.)

Like the knot group, the genus has excellent potential as a way of determining whether a given knot is really knotted, or not. The genus of the trivial knot is obviously zero, since the circle bounds a disc with no handles. Moreover, though not so obvious, the trivial knot is the only one with genus 0. So in order to show that a knot is really a knot, all you need to do is show that its genus is non-zero. But how do you go about calculating the genus of a knot – given, say, a diagram of the knot (and that is what you are usually presented with)? A possible method was suggested in 1962 by Wolfgang Haken (of four-colour-theorem fame – see Chapter 7), and more recently, in 1978, Geoffrey Hemion used Haken's results to construct an algorithm which will always decide whether two given knot diagrams represent equivalent knots. Unfortunately the algorithm is far too inefficient to be of much practical use (see Chapter 11 for a discussion of the efficiency of algorithms), but it does show that the problem of distinguishing between knots is, in principle, one that can be decided in a mechanical fashion.

Incidentally, though Seifert's construction was developed geometrically rather than algebraically, in 1978 Charles Feustel and Wilbur Whitten showed how to obtain the genus from the knot group, thereby emphasizing once again the fundamental nature of the knot group.

By this stage you, the reader, are no doubt beginning to reel at the amount of heavy mathematical machinery that has been brought into play in order to distinguish between knots. 'Is there not a simpler way?' you might ask. Well, provided you are prepared to accept an occasional failure to distinguish between non-equivalent knots, then there is indeed a simpler way. It is provided by the so-called *knot polynomials*, the simplest of which are the *Alexander polynomials*, discovered by J. W. Alexander in 1928. Though it may be derived from the knot group, the Alexander polynomial of a knot can be obtained directly from its diagram. For the trivial knot the Alexander polynomial is the number 1. For the trefoil (Figure 53(iii)) it is

$$x^2 - x + 1,$$

and for the four-knot (Figure 53(iv)) it is

$$x^2 - 3x + 1.$$

Since these two polynomials are obviously not the same, it follows at once that the trefoil and the four-knot are not equivalent (and that neither of them is trivial). The Alexander polynomial of a knot is a very useful invariant: it is adequate to distinguish between all prime knots with crossing number up to 8, and all but six pairs with crossing number 9.

But one pair of knots which the Alexander polynomials cannot tell apart are the reef knot and the granny knot. These both have crossing number 6, but they are not prime, each being composed of two trefoils. The difference between them is whether the trefoils are left- or right-handed (see Figure 55). Both knots have the Alexander polynomial

$$(x^2 - x + 1)^2.$$

(It is no accident that this is just the square of the Alexander polynomial of the trefoil. The Alexander polynomial of any composite knot is just the product of the Alexander polynomials of the constituent knots.) But don't blame the Alexander polynomial for this failure. As mentioned earlier, the much more powerful knot group cannot distinguish between them. The problem is in differentiating between a left-handed trefoil and a right-handed trefoil. The knot group cannot do this, at least not on its own. By 'enhancing' the knot group in a suitable fashion it is possible to obtain a knot invariant which will do the job, but once again the solution involves still more abstract machinery.

So is it the case that what the average Boy Scout can do with ease has the mathematician searching through his toolbag of highly sophisticated, abstract apparatus? Until August 1984 the answer was yes. But then

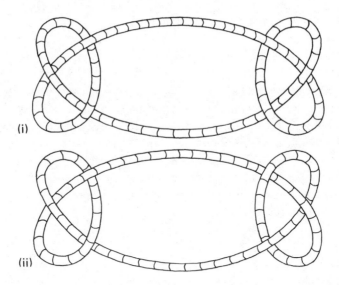

Figure 55. (i) Reef knot, (ii) granny knot. The difference between the two is the way the right-hand trefoil winds around in the two knots. The knot group cannot distinguish between these two orientations.

things brightened up considerably with the simultaneous discovery by *four* quite independent groups of mathematicians (one in Britain, the others in the USA) of a new type of polynomial which can indeed tell a left-handed trefoil from a right-handed one (and thus a reef knot from a granny). Because of the manner in which the new polynomials were discovered, naming them after those who had done the work would mean calling them the Conway–Jones–Freyd–Yetter–Hoste–Lickorish–Millett–Ocneanu polynomials, which is quite a mouthful. Fortunately, the polynomials themselves are much simpler. Whereas the Alexander polynomials have just one variable, the new polynomials have two. (They also involve negative powers of these variables, so the reader might at first baulk at calling them 'polynomials'.) The new polynomial for the right-handed trefoil is

$$-x^{-4} + x^{-2}y + x^{-2}y^{-1}$$

and the one for the left-handed trefoil is

$$-x^4 + x^2y + x^2y^{-1}.$$

For the four-knot, the new polynomial is

$$x^{-2} + x^2 - y - y^{-1} + 1.$$

How do you calculate these polynomials? Well, you obtain them from the knot diagram. Though not within the scope of a book such as this, the procedure is sufficiently mechanical to be performed on a computer in an acceptably efficient manner.

Do the new polynomials serve to distinguish between all non-equivalent knots? No, they do not. For that you must keep on searching. In the theory of knots there is still a lot to be unravelled, but for us it is time to move on to something else.

Scratching the Surface

The central thrust in the development of topology since the mid 1950s has undoubtedly been in the study of *manifolds*. Loosely speaking (and a proper definition will be given later), a manifold is a generalization of the notion of a surface to any number of dimensions. So the simplest kinds of manifold are the one-dimensional ones, which are just curves (with the real line, $\mathbb{R}$, a special case) and the two-dimensional ones, the surfaces (with the flat two-dimensional plane, $\mathbb{R}^2$, a special case). Two-dimensional manifolds have the advantage that it is possible to draw pictures of them (or, even better, make physical models of them). Unfortunately for the readers of this book, practically all the work on manifolds that has been done since the turn of the century has concerned manifolds of dimension three or more, and such manifolds cannot be properly illustrated. (Diagrams in books dealing with higher-dimensional manifolds are always meant for the expert, and require great care in their interpretation.) Of course, if the dimension is not too high, say three or four, then it is possible to illustrate simple manifolds using projections or cross-sections. But again, these usually require some accompanying explanation if they are to be understood. For instance, without the caption would you have recognized the four-dimensional object that Figure 56 is intended to illustrate? What are shown are two projections of a *hypercube*, a four-dimensional analogue of a cube. Just as a three-dimensional object can be illustrated on a two-dimensional page by means of a projection, so too – in principle – can

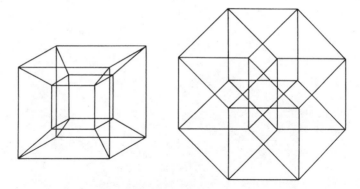

Figure 56. The hypercube, a four-dimensional figure having eight equal-sized cubes as its 'faces'. Both illustrations are flat projections of a (three-dimensional) 'projection' of a hypercube into three-dimensional space.

a four-dimensional object be depicted by projecting it down to a three-dimensional object in space. By further projecting this three-dimensional object onto a page, it is possible to get an illustration of the original four-dimensional object, and this is what you see in Figure 56. Of course, some considerable mental effort is required to interpret such a projection. With flat projections of three-dimensional objects, the human mind is quite good at doing this. Indeed, some of the disturbing art of the Dutch painter Mauritz Escher achieves its effect by exploiting this facility in order to create 'impossible' figures like the one shown in Figure 57. Interpreting a projection of a four-dimensional object is more difficult – what does a hypercube 'look like'? Well, just as an ordinary cube has six faces, each of which is a square and all of which are the same size, so a hypercube has eight 'faces', each of which is a cube and all of which are the same size. If you look carefully at the left-hand diagram in Figure 56 (and try to visualize the three-dimensional projection that it depicts) you will see that it shows eight cubes: a large outer one, a smaller inner one, and six which are distorted into the form of truncated pyramids. The distortions in size and shape are all features of the initial projection from four down to three dimensions. In (four-dimensional) 'reality', all eight cubic 'faces' are the same size, and the hypercube is what lies 'between' them.

Besides projections, another way of trying to visualize higher-dimensional objects is to take a succession of cross-sections. For instance, what four-dimensional object will give the sequence of three-dimensional cross-sections

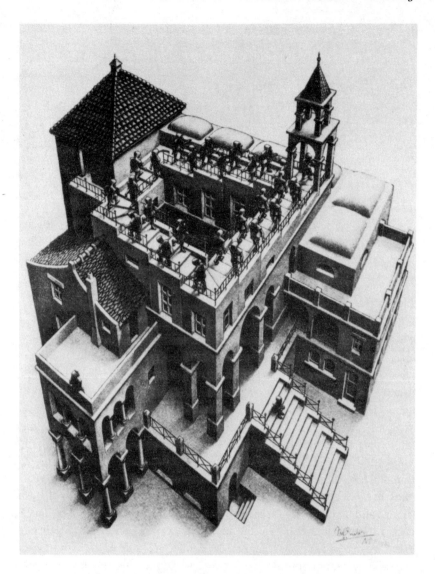

Figure 57. The art of M.C. Escher: *Ascending and Descending* (1960). The human facility for interpreting a three-dimensional object from its two-dimensional projection is exploited in order to achieve an impossible figure.

shown in Figure 58? Well, successive two-dimensional cross-sections of a sphere will form a sequence of circles, starting from very small ones, growing to a maximum, then growing smaller again. (This will be familiar to anyone who has ever sliced an apple.) In the same way, successive cross-sections of a four-dimensional *hypersphere* will form a sequence of spheres, as shown in Figure 58. (Question: What will a sequence of cross-sections of a hypercube look like?)

Figure 58.   Cross-sections of a four-dimensional hypersphere. This is what you would obtain if you were able to slice up a four-dimensional apple: a sequence of spherical 'slices' which grows to a maximum size and then decreases to nothing again.

Such illustrative devices can give you at best a vague idea of what a four-dimensional object 'looks like'. They are no use whatsoever in trying to appreciate objects of five or more dimensions. But what they do very well is to indicate how the progression from lower to higher dimensions proceeds in a step-by-step fashion of analogy. For instance, the bounding 'faces' of a two-dimensional square are four equal-sized one-dimensional straight lines, the bounding faces of a three-dimensional cube are six equal-sized two-dimensional squares, the bounding 'faces' of a four-dimensional hypercube are eight equal-sized three-dimensional cubes, and so on.

But beware! All is not as might at first appear. In three-dimensional space there are, as the Ancient Greek geometers knew, just five regular polyhedra: the tetrahedron, cube, octahedron, dodecahedron, and icosahedron. In more than three dimensions the analogue of a polyhedron is called a *polytope*. (This concept is also mentioned in Chapter 11, along with an application in the real, three-dimensional world.) The 'faces' of a four-dimensional poly-tope are three-dimensional polyhedra. For a regular polytope these 'facing' polyhedra must themselves be regular, and the arrangement of the 'faces' must be the same at each vertex. It turns out that there are just six regular four-dimensional polytopes: the *simplex*, having as 'faces' five tetrahedra, the *hypercube*, with eight cubes as 'faces', the *16-cell*, bounded by 16 tetrahedra, the *24-cell* with 24 octahedra, the *120-cell* with 120 dodecahedra, and the *600-cell*, with 600 'faces', each a tetrahedron. So things get a little more complicated as you go up from three to four dimensions. But then something curious happens. For any number of dimensions greater than four, there are only three regular polytopes, analogous to the tetrahedron, cube, and octahedron. So what is going on? Why should things suddenly become simpler (and constant) beyond four dimensions? Though no one really knows the answer to this question, the phenomenon is not one that is restricted to regular polytopes. In many other respects also, spaces of five or more dimensions are much easier to deal with than three- and four-dimensional space.

But, despite the fact that some quite new factors come into play in higher dimensions, looking at what goes on for two-dimensional manifolds (surfaces) still provides a reasonable idea of the kind of problems and methods that are involved in manifold theory, and consequently it is worth taking a closer look at the topological theory of surfaces.

The classification of all two-dimensional manifolds, mentioned earlier, was one of the great triumphs of nineteenth-century topology. Just two invariants are required to distinguish between all closed surfaces: orientability and the Euler characteristic. How can this classification be achieved? Modern proofs usually proceed in two stages. First of all it is shown that every closed surface can be topologically deformed into one of two standard forms. Then all that remains to be done is to show that the two invariants, orientability and Euler characteristic, are enough to distinguish between all standard surfaces.

The *standard orientable surface of genus n* consists of a sphere to which are attached *n handles*. To attach a handle to a surface, you cut two holes in the surface and then sew in a cylindrical tube to join the two holes together (see Figure 59). A sphere with any number of handles is an orientable surface. The Euler characteristic of a sphere with n handles is $2 - 2n$. This is not difficult to prove – the trick is to start with a network on a sphere (for which $V - E + F = 2$) and then add handles to it. If this is done carefully enough you will see that, each time a handle is added, the Euler characteristic decreases by 2.*

By means of a process consisting of cutting, pulling apart, and reassembly, known – for obvious reasons – as *surgery*, it is not at all difficult to deform any given orientable surface into a standard orientable surface of some genus. For instance, a torus will give one of genus 1, a double-torus will give one of genus 2, and so on. Since the genus and the Euler characteristic for the standard orientable surfaces are related (by the above expression

---

*Full details of this procedure, and indeed the entire classification proof, can be found in the book *Concepts of Modern Mathematics*, by Ian Stewart (Penguin, 1981), Chapter 12.

$2 - 2n$), this shows how the Euler characteristic serves to classify all orientable surfaces.

The *standard non-orientable surface of genus n* is obtained by taking a sphere and adding to it *n cross-caps*. To add a cross-cap to a surface, you cut a hole in the surface and sew a Möbius band across it, edge-to-edge. In three-dimensional space this can be achieved only if you allow the Möbius band to intersect itself (see Figure 60). Because a Möbius band enables clockwise and anticlockwise directions to be interchanged, a surface with a cross-cap will be non-orientable. The Euler characteristic of a sphere with n cross-caps is $2 - n$. Again, this can be established by starting with a network on a sphere and observing that each addition of a cross-cap decreases the quantity $V - E + F$ by 1. (This is fully explained in Stewart's book, mentioned above.)

By using surgery it is possible (and not particularly difficult) to deform any non-orientable surface into a standard non-orientable surface of some genus. For instance, the geometer's *projective plane* (which, despite its

Figure 59. Handles. To attach a 'handle' to a surface, cut two holes, as in (i), and sew a cylindrical tube to the two edges to connect them together, as in (ii). Using a technique known as surgery it is possible to show that every orientable closed surface is topologically equivalent to a sphere to which are attached a certain number of handles. This gives a 'standard form' for orientable closed surfaces.

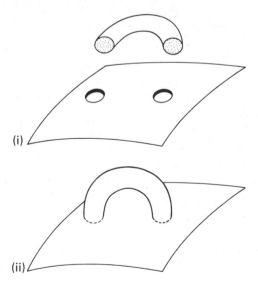

name, is a closed surface) will transform to a standard non-orientable surface of genus 1, and the Klein bottle will transform to one of genus 2. Since the Euler characteristic of a standard non-orientable surface is related to the genus (by the expression $2 - n$), this standardization procedure establishes the classification by the Euler characteristic of all non-orientable surfaces as well. Surfaces with edges are dealt with by allowing the standard surfaces to have holes in them.

With the above knowledge of surface topology as a jumping-off point, we can go on to see what has been happening in higher dimensions in recent years. We start by looking at one of the simplest types of manifold: the n-dimensional sphere, for n equal to 2, 3, 4, and beyond. It is with these manifolds that the most famous problem in topology is concerned.

Figure 60. The cross-cap. To attach a cross-cap to a surface, cut a hole, as in (i), and sew a Möbius band across it edge-to-edge. In three-dimensional space this can be visualized only if the Möbius band is allowed to intersect itself, as in (ii). The resulting sewn-in piece is the cross-cap. By means of topological surgery it is possible to show that any non-orientable closed surface is topologically equivalent to a sphere with a certain number of cross-caps. This gives a 'standard form' for non-orientable closed surfaces.

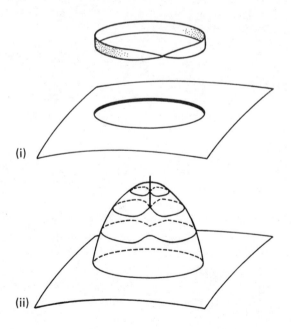

The Poincaré Conjecture

The simplest of all closed two-dimensional surfaces is the sphere, the surface which acted as the starting point for the classification process described above. The *n*-dimensional analogue of a sphere is known as an *n-sphere*. (So the ordinary sphere is a 2-sphere.) Just as the 2-sphere is the surface of a three-dimensional solid ball, so the *n*-sphere is the 'surface' of an $(n + 1)$-dimensional 'solid ball'. When the French mathematician Henri Poincaré began to investigate higher-dimensional manifolds at the beginning of this century (and thereby essentially started the subject of manifold topology as it is understood today), naturally enough he paid particular attention to the *n*-spheres. They ought, after all, to be very special, just as the 2-sphere is special amongst two-dimensional manifolds. In 1904, unable to prove what he thought was an extremely reasonable assertion about *n*-spheres, he formulated the assertion as a conjecture which was destined to become the most famous problem in the field. Like all good conjectures, it is both fundamental and easy to state.

Suppose you draw a closed loop on a 2-sphere. Then it is possible to shrink the loop down to a point without the loop leaving the surface (see Figure 61(i)). Moreover, the sphere is the only closed surface for which this is possible. If you were to draw your loop on a torus, for example, in either of the ways indicated in Figure 61(ii), then it could not be shrunk down to a point. Similarly (only now you have to rely on abstract mathematics without the aid of pictures), if you take an *n*-sphere for any value of *n* greater than 2, and 'draw' a loop on it, then that loop may also be shrunk down to a point without leaving the *n*-sphere. But (and this is the big question) is the *n*-sphere the only *n*-dimensional closed manifold which has this property, as it is in two dimensions? The *Poincaré conjecture* says that the answer is 'Yes'. (Strictly speaking, Poincaré was interested only in three-dimensional manifolds.)

Despite a great deal of effort, the problem of proving (or disproving) the Poincaré conjecture resisted all attempts at solution until 1960, when the American mathematician Stephen Smale proved that the conjecture is valid for all dimensions from five upwards. This result was sufficiently highly regarded to merit Smale being awarded a Fields Medal for his work. It also provided an instance of the phenomenon mentioned earlier, that manifolds behave differently from dimension five upwards: Smale's methods did not work for three or four dimensions. In fact it was almost twenty years before the problem was solved for four dimensions, also by an American. In 1981, Michael Freedman built upon Smale's ideas and work by Andrew Casson in order to prove the Poincaré conjecture for 4-spheres. Freedman's result followed as a special case of a much more general result on four-dimensional manifolds, of which more in the next section.

That left the three-dimensional problem, the one for which the conjecture was originally formulated. What is the present state of affairs? Despite a great deal of work by first-rate mathematicians, it remains unsolved. Some measure of the difficulty encountered in trying to solve the problem can be gleaned from the fact that it took experts several months to find a flaw in one of the most recent attempts at a solution, announced early in 1986. Eighty years after it was put forward, the Poincaré conjecture is still not ready to give up its status as *the* unsolved problem in topology.

Figure 61.   The Poincaré conjecture. If a closed loop is drawn on a sphere, it is possible to shrink that loop to a point without it leaving the surface, as in (i). On a torus this is not always possible. If the loop is drawn in either of the two ways indicated in (ii), it cannot be shrunk to a point without leaving the surface. For closed surfaces, this property of being able to shrink any closed loop to a point is totally characteristic of the sphere: no other closed surface has this property. The Poincaré conjecture says that an analogous result holds for all higher dimensions. For instance, the only three-dimensional closed manifold with the loop-shrinking property is the three-dimensional hypersphere.

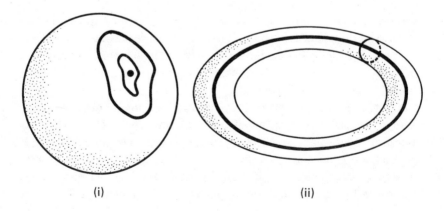

(i)   (ii)

The Theory of Manifolds

The abstract definition of an n-dimensional manifold is that it is an object with the property that if you look at any small part of it, what you see looks very much like ordinary (?) n-dimensional Euclidean space, $\mathbb{R}^n$. For example, the 2-sphere is a two-dimensional manifold according to this definition. Any small part of the sphere does indeed look like $\mathbb{R}^2$ – a fact which is apparent to us as we walk around on the surface of the Earth. Note that because an n-dimensional manifold has to look like $\mathbb{R}^n$ on a *local* basis, it does not follow that the entire manifold looks like $\mathbb{R}^n$. It was the failure to appreciate this (for two-dimensional manifolds) that led to the belief that the Earth is flat. Locally it is (almost) flat; globally it is not.

Now, the main interest in the study of manifolds lies in the natural way they arise in connection with problems in analysis and physics, and in these cases you get more than a bare manifold as defined above. What you get is a manifold on which it is possible to develop one of mathematics' most powerful techniques: the differential calculus (i.e. you can 'do differentiation' on it). You are probably familiar with the way the differential calculus is developed on the manifold $\mathbb{R}$. This is the 'ordinary' differential calculus of functions of one real variable, taught in schools. You may even have seen how to do it on the manifold $\mathbb{R}^2$ as well. And the same kind of techniques will allow you to develop a differential calculus for any of the higher-dimensional manifolds $\mathbb{R}^n$, for $n = 3, 4, 5$, and so on.

Since any n-dimensional manifold is locally like $\mathbb{R}^n$, you can of course make use of the methods of the differential calculus on that manifold *on a local basis*. But what about globally? Well, for the sphere at least it is possible to develop a differential calculus that covers the entire surface. The reason for this is that the transition from one localized area (which looks like $\mathbb{R}^2$) to the next is an essentially smooth and trouble-free one. To put it another way, suppose you were to cover the entire surface of the sphere by lines of latitude and longitude. Then you would get a coordinate system which, on a local basis, looks just like the usual coordinate system of Cartesian geometry (which is what lies behind the development of the differential calculus on $\mathbb{R}^2$, of course). If you use these coordinates in order to develop your calculus locally on the surface, then because you are using the very same coordinate lines all over the surface, there will be no conflict between what happens at one location and what goes on at another: the transitions will all be *smooth* ones.

So a natural and fundamental question is for how many manifolds it is possible to develop a differential calculus covering the entire manifold, as can be done for the sphere. A manifold for which it is possible to develop a global theory of differentiation is called a *smooth* (or sometimes a *differentiable*) *manifold*. A coordinate system which covers the entire manifold and serves as a basis for the differentiation process (such as the lines of latitude and longitude on the sphere) is called a *differentiation structure*. (Actually things are a bit more complicated, but this is more or less the picture.) The basic question of which manifolds are smooth (or which manifolds can be given a differentiation structure) carries with it the no less interesting ancillary question of whether, given a smooth manifold, there is only one differentiation structure which can be given to the manifold, or whether it can be done in more than one way (and if so, in how many ways). Moreover, since physicists spend a great deal of their time working with the calculus on various manifolds, the answers to these questions will be of interest not just to the topologists.

For manifolds of two or three dimensions, the answers to the above questions were known by the mid 1950s: every two- or three-dimensional manifold is smooth, and no such manifold can be given two essentially different differentiation structures. It was, it seemed then, only a matter of time before the result would be extended to cover manifolds of all dimensions. But in 1956, John Milnor of the USA discovered, much to everyone's surprise, that the 7-sphere can be given 28 quite distinct differentiation structures, and soon afterwards it was found that other higher-dimensional spheres could be given more than one differentiation structure. Obviously there was a great deal of work to be done in order to try and see what was going on, and there was no shortage of able mathematicians ready to do that work. The period from 1956 to 1970 has been called a 'golden era of manifold topology'. (What it was, in fact, was a golden era of the study of manifolds of dimension five or more, for once again the problem for four dimensions proved intractable by the methods available.) During this period, by using a mathematical concept called homotopy, topologists were able to obtain a fairly systematic classification of all manifolds of dimension greater than four, distinguishing in particular between the smooth and the non-smooth ones.

But what then of four dimensions? Would all manifolds be differentiable and allow only a single differentiation structure, as with lower dimensions? Or would there be a whole range of possibilities requiring classification, as with higher dimensions? The answer finally came in 1981. Michael Freedman, besides resolving the Poincaré conjecture for four dimensions (see earlier), also established that there is a four-dimensional manifold which is not smooth. (Freedman's description of this manifold – which for technical reasons is known as $\mathbb{E}_8$ – is, like everything else in higher-dimensional topology, an algebraic one.) In fact both the four-dimensional Poincaré conjecture and the non-smooth four-dimensional manifold result followed from a single, very general (and totally unexpected) result that Freedman obtained, which showed that just two 'elementary' pieces of information are all that is required to classify any four-dimensional manifold. (These pieces of information are not so 'elementary' that they can be explained here, however!)

But that was not the end of the story. There was still another, equally dramatic surprise in store for the topologists, and it was not long in coming. And this surprise was one that hits right at the heart of the very physical universe we live in!

Unexpected results about manifolds can always be explained away on the grounds that one is, after all, dealing with some pretty abstract notions that can be at best only partly appreciated. Even two-dimensional manifolds can be pretty fancy objects, all but defying the imagination. The same surely cannot be said of the 'concrete' manifolds $\mathbb{R}$, $\mathbb{R}^2$, $\mathbb{R}^3$, and so on. After all, $\mathbb{R}^3$ is the physical space we live in (isn't it?), and $\mathbb{R}^4$ is the space–time continuum. And indeed these 'concrete' manifolds do exhibit almost exemplary behaviour. For a start, they are all smooth. Moreover, for each n there is just one way of assigning a differentiation structure to $\mathbb{R}^n$ – except, that is, for the one case $n = 4$.

For some peculiar reason, mathematicians had been unable to find a proof of the uniqueness of a differentiation structure that worked for $\mathbb{R}^4$. Every other $\mathbb{R}^n$, yes, but not $\mathbb{R}^4$. What made this lack of success all the more embarrassing was that this was the very case of greatest interest to the physicists. Still, it could only be a matter of time before a proof was found. It was inconceivable that there could be a non-standard way of doing differentiation on $\mathbb{R}^4$ ... wasn't it?

But then the inconceivable turned out to be true. The bombshell was dropped in the summer of 1982. Combining Freedman's work (which is essentially algebraic) with a fairly hefty dose of analysis and differential geometry, Simon Donaldson, a 24-year-old student of Michael Atiyah at Oxford University, proved a result that implied the existence of a differentiation structure on $\mathbb{R}^4$ other than the usual one. In other words, the differentiation structure used by physicists and mathematicians the world over is not unique! (Indeed, subsequent work by Clifford Taubes showed that the usual differentiation structure on $\mathbb{R}^4$ is just one of *infinitely many* that may be given to this manifold!) This raises two intriguing questions. What is so special about four dimensions which makes this phenomenon occur only in this one case? And as there is more than one way of performing differentiation on $\mathbb{R}^4$, how do we know which is the 'right' one as far as physics is concerned? With *n*-dimensional manifolds beginning to fall into order for all values of *n* other than 4, the case $n = 4$ has become curiouser and curiouser.

So, are physicists working with the right mathematics on $\mathbb{R}^4$? Well yes, they probably are. The infinitely many 'exotic' differentiation structures on $\mathbb{R}^4$ that have been discovered all involve some very special 'cooked-up' behaviour which rules them out as far as our own physical universe is concerned. But their existence does indicate in a very striking manner that there is something very special indeed about four-dimensional space. And their discovery shows that this is indeed a Golden Age as far as topology is concerned.

Figure 62.   Solution to the ring puzzle (Figure 48). The sequence shown indicates how the original linked ring configuration can be deformed to an unlinked ring figure.

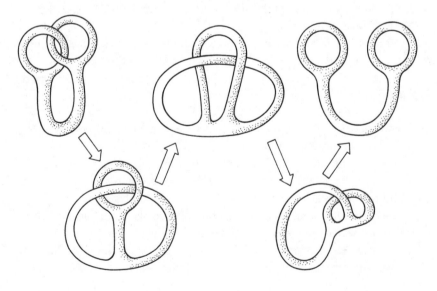

Suggested Further Reading

For a more complete introduction to topology than could be given here, though at essentially the same level, see *Concepts of Modern Mathematics*, by Ian Stewart (Penguin, 1981), Chapters 10–14.

A layman's introduction to knot theory can be found in the article 'The theory of knots', by Lee Neuwirth, which appeared in the magazine *Scientific American*, Volume 240 (June 1979), pp. 84–96. A more complete coverage (at a higher level) is given in the book *Introduction to Knot Theory*, by Richard Crowell and Ralph Fox (Gimm and Company, USA, 1963). And a nice survey of the field is given in an article entitled 'Knot tabulations and related topics', by Morwen Thistlethwaite, which appeared in the compilation volume *Aspects of Topology*, edited by I. M. James and E. H. Kronheimer, London Mathematical Society Lecture Note Series, Volume 93 (Cambridge University Press, 1985).

Other than at the level of, say, Stewart's book mentioned above, manifold theory tends to be pretty impenetrable to the outsider. To get some idea of what it is like you could take a look at the book *Instantons and Four-Manifolds*, by Daniel Freed and Karen Uhlenbeck (Springer-Verlag, 1984), which concentrates on the recent work by Donaldson and others on non-standard $\mathbb{R}^4$ manifolds. (But only think of actually buying this book if you are already an expert on algebraic topology. If you are not, it is not for you.)

11 The Efficiency
of Algorithms

Algorithms Again

The concept of an algorithm has already played a prominent role in this book, in Chapter 6. (And it will be assumed from now on that you have read that chapter.) With Hilbert's tenth problem (considered there) the issue was whether or not a particular problem could be solved by an algorithm, with the stress on the 'could be'. Pure existence was the order of the day – there was no question about whether the algorithms being discussed were at all practicable. In the context of Hilbert's question, this was, of course, perfectly in order. But when it comes to problems that arise in the real world around us, the mere existence of an algorithm is by no means the end of the matter. Indeed, it is just the start, as there is no real advantage to be gained from having an algorithm which, though theoretically capable of solving the problem at hand, might require thousands of years of high-speed computer time to do so. For the kinds of problem that are of concern to businessmen and applied scientists, the important issue is the existence of an efficient algorithm. In a business application this might mean the ability to obtain a solution within a few hours. For something like an aircraft-guidance system, results need to be available within a fraction of a second. For applications such as these, it is clearly useful to be able to prove that a particular problem can or cannot be solved by means of an 'efficient' algorithm. In order to do this, the first step is to formulate a suitable method for evaluating algorithm efficiency.

Obviously, the speed with which a given task can be performed on a computer depends on a number of factors. The size and operating speed of the computer, the efficiency of the programming language used to write the program, and the skill of the programmer are all relevant. But these are highly specific factors which are not suited for a general study. What we require is some very general way of dividing algorithms into two categories: efficient and non-efficient. This classification should be sufficiently robust that altering such 'peripheral' factors as the computer speed or the programming language will not convert an inefficient algorithm into an efficient one, or vice versa.

Such an efficiency classification was introduced by A. Cobham and J. Edwards in the mid 1960s, and now forms the basis for most work on algorithm efficiency. Though their fundamental measure of efficiency is referred to in terms of 'time', in order to avoid a dependence of computing speeds the actual definition is given in terms of the number of steps required to perform the calculation. Of course, even this notion is not absolute – it depends upon what constitutes a basic step and on how the data is represented. But it turns out that all such considerations are irrelevant as far as the fundamental notion of efficiency is concerned. For this reason it has become standard to formulate the various definitions in terms of Turing machines (see Chapter 6). These are simple enough to allow for a smooth mathematical theory, and yet all the results obtained are equally valid if you recast everything in terms of your favourite computing device, whatever it may be.

Having decided on the Turing machine as the basic computing device for the theory, the idea is to measure the efficiency of an algorithm by the number of steps (i.e. Turing-machine steps) taken to complete the calculation. Questions about how the algorithm is coded as a Turing-machine program and how the data is coded on the tape turn out to be of no significance (that is to say, such considerations do not affect the boundary between efficient and inefficient algorithms). What is relevant is the size of the data that has to be handled. The more data there is, the more steps are required to handle it. For example, if you are multiplying pairs of integers by hand, it will take you just over four times as long when you move up to numbers twice the length of the previous pair: there are four times as many basic digit multiplications, plus the associated 'overheads' of keeping track of the various carries. Bearing this in mind, here then are the basic definitions.

An algorithm (a Turing-machine program for the purpose of this definition) is said to run in *polynomial time* if there are fixed integers A and k such that for input data of length n, the computation is completed in at most An^k steps (for any value of n).

For example, the standard algorithm for adding (by hand) two whole numbers runs in polynomial time. If the numbers are expressed in standard decimal format and the basic computational operation is the addition of two digits, then the addition of two numbers each with $n/2$ digits (input data of length n) involves exactly n steps (allowing for carries), so the above definition is satisfied with both A and k equal to 1. In the multiplication of two $n/2$-digit numbers there are $n^2/4$ basic digit multiplications plus $n/2$ carries, giving $n^2/4 + n/2$ steps in all. Since $n^2/4 + n/2$ is always less than n^2, if you take $A = 1$ and $k = 2$ in the definition you will see that integer multiplication (by the standard method) is a polynomial-time algorithm.

If the above examples are given in terms of Turing machines rather than decimal digit arithmetic, you would of course need to use larger values of the constant A, and possibly even a larger k as well, but you would still be dealing with a polynomial-time algorithm. In fact this is why the notion of a polynomial-time algorithm is independent of any changes in machine and programming details. These lead only to changes in the size of the two constants; the definition remains valid.

Algorithms which do not run in polynomial time are said to run in *exponential time*. For example, an algorithm which requires 2^n (or 3^n, or n^n, or $n!$) steps to handle input data of length n is an exponential-time algorithm. This explains the use of the word 'exponential' here, though the usage is a bit misleading since it includes functions like $n^{\log n}$, which is not usually regarded as an 'exponential' function.

As you will have realized by now, 'efficient' algorithms are ones that run in polynomial time – and 'inefficient' ones are the ones which require exponential time. The discussion of exponential growth in Chapter 1 should be enough to convince you that exponential-time algorithms might well be highly inefficient (but see later), but you might well be sceptical about whether polynomial-time algorithms are necessarily efficient. The freedom to choose the constants A and k in the definition of polynomial time would seem to provide far too much leeway: an algorithm which is classified as 'efficient' only when you choose $A = 10^{10}$ and $k = 100$ is hardly likely to be 'efficient' in any real sense. Two points are worth making here. Firstly, what happens in practice is that problems turn out to be solvable either by exponential-time algorithms only, or by polynomial-time algorithms of the order of, say, $10n^3$ steps, or perhaps even less. Secondly, the polynomial/ exponential split is only a crude, preliminary classification. In the future it may well be necessary to look for a more sensitive distinction, but at the moment the one available works quite well. Both these remarks are emphasized to great effect in Table 4.

Time-complexity function	Size of data: n					
	10	20	30	40	50	60
n	0·00001 s	0·00002 s	0·00003 s	0·00004 s	0·00005 s	0·00006 s
n^2	0·0001 s	0·0004 s	0·0009 s	0·0016 s	0·0025 s	0·0036 s
n^3	0·001 s	0·008 s	0·027 s	0·064 s	0·125 s	0·216 s
2^n	0·001 s	1·0 s	17·9 min	12·7 days	35·7 years	366 centuries
3^n	0·059 s	58 min	6·5 years	3855 centuries	2×10^8 centuries	$1·3 \times 10^{13}$ centuries

Table 4. Polynomial and exponential running times. It is assumed that a computing device performs one basic operation in 0·000 001 seconds. For a given size of data and 'time-complexity function' (i.e. how the computation depends on the data size), the table gives the time required to perform the computation. Notice the much more explosive growth rates for the two exponential functions. The computation time for $n = 50$ and a time-complexity function 3^n exceeds the best current estimates of the age of the Universe, and for $n = 60$ it is about 100 000 times longer even than that!

As an illustration of the way the above notions are used to classify real problems and algorithms, we now consider a famous problem whose importance in the commercial and business world is self-evident.

The Travelling-Salesman Problem

Imagine you are a travelling salesman and that you have to visit a number, say 50, of different locations. The order in which you make your visits is not important, as long as you get to them all. It is obviously in the interest of both yourself and whoever is paying your transport costs that you make your visits in the order which requires the least amount of travelling (in total). How do you go about choosing your route? Obviously

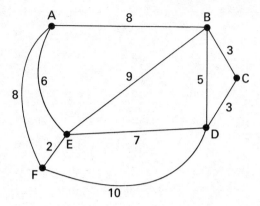

Figure 63. The travelling-salesman problem. Find a tour of all the locations shown that minimizes the total distance travelled. (The location-to-location distances along the available paths are indicated.) For instance, the tour ABEFDC means a journey of length $8 + 9 + 2 + 10 + 3 = 32$. (Sometimes it is required that the tour starts and finishes at the same location, in which case that particular route would be incomplete.)

you start by drawing up a table showing the distance between each pair of locations to be visited. But having done that what do you do? For example, what is the most economical route for visiting all the locations in Figure 63?

An obvious approach is to list *all* possible routes, work out the total distance for each, and pick the one which gives the smallest answer. This will certainly work, which demonstrates that the problem can be solved by means of an algorithm since the procedure could easily be performed by a computer, at least in principle. But even for quite modest numbers of locations the number of possible routes is far too great to handle. If there are N locations to visit, then there are $N!$ possible itineraries. (Recall that $N!$, read as 'N-factorial', is the product of all the numbers N, $N - 1$, $N - 2$, down to 3, 2, and 1.) Since the function $N!$ is certainly exponential (it grows faster than 2^N or 3^N, though not as fast as the 'superexponential' function N^N), listing all possible routes will obviously lead to an exponential-time algorithm. To see just how badly this method will fare, note that for 10 locations there are

$$10! = 3\,628\,800$$

possible routes. This could be handled on a modern computer, but when we get up to 25 locations, the number of routes to consider is a daunting 16 followed by 25 zeros. And a tour of 25 cities is not at all unrealistic for a real-life salesman, to say nothing of other situations involving essentially the same mathematical problem where the number of 'locations' might be in the hundreds.

So arriving at a solution by listing all the possibilities is obviously out of the question except when only a small number of locations are involved. What else might you try? A 'common-sense' solution maybe? For instance, by looking at the map (or table of distances) you could work out a route which involves first visiting all locations close to your starting point, and then moving further afield. Though this strategy (and any other you might like to try) may well work in certain special situations, it has been shown that it will not work in all cases. And it is the overall behaviour of the algorithm that concerns us now. Individual instances of the salesman's problem might turn out to have simple solutions (for example, if all the locations lie on a straight line the shortest route is obvious), but what we want is an algorithm that will work in all cases. However, despite the great deal of effort that has been put into the problem (over 300 technical papers have been written on it) since it was first raised by the Viennese mathematician Karl Menger in 1930, the travelling-salesman problem has resisted all attempts at a general solution. In fact, as we shall see, there is very strong evidence that there is no efficient algorithm for this problem!

In the meantime, some mention should however be made of the considerable advances that have been achieved for some special cases. For a start, in 1962 Michael Held and Richard Karp of IBM used a technique called dynamic programming to solve the problem for all tours of at most 13 locations ($13! = 622\,702\,080$). In 1963 Little, Murty, Sweeney, and Karp invented a powerful technique called 'branch and bound' which enabled the problem to be solved for tours of up to 40 locations in just a few minutes of computer time on a mainframe computer (an IBM 7090). In 1970, Held and Karp developed a branch-and-bound algorithm which was able to solve a specific instance of the problem, with 42 locations. This algorithm required the examination of only 61 of the vast number (33 followed by 49 zeros) of possible routes. (The significance of this particular problem, which involved 42 cities spread across the USA, was that a solution had been obtained in 1954 in Dantzig, Fulkeston, and Johnson of the RAND Corporation.) And in 1979 Crowder and Padberg solved a specific problem involving 318 locations, at the time by far the largest number ever handled. The current position is that the techniques available should be adequate for the solution – in a reasonable amount of computing time, say a few days at the most – of any instance involving up to about 500 locations. But the structure of the individual instance is a crucial factor. By and large, problems that arise from real life – actual tours between cities – turn out to be amenable to solution, whereas it is possible to devise 'artificial' instances that will defeat all known methods of attack. The next section indicates why a general solution which works well in all cases is very likely not possible.

P and NP

For an abstract discussion of how amenable problems are to solution by efficient algorithms, it turns out to be convenient to reformulate all problems as ones which require a simple yes/no answer; this allows different problems to be compared. For example, the multiplication problem (Given integers a and b, what is their product?) could be recast as: Given integers a, b, and c, does $ab = c$? The travelling-salesman problem could be recast as: Given a collection of locations, together with a table of the distances between them, and given a number B, is there a tour of the locations whose total travelling distance is at most B? (It is not immediately apparent that this entirely captures the essence of the problem as originally formulated, but in fact it does. If there is an efficient algorithm for solving the original version, then there is one for solving the simplified version, and vice versa.)

Problems for which yes/no answers are required are called *decision problems*. A decision problem is said to be of type P if it can be solved by an algorithm which runs in polynomial time. For example, the multiplication problem mentioned above is of type P. To see if $ab = c$, simply multiply a and b together and see if the result is equal to c. This requires only polynomial (in fact quadratic) time.

The travelling-salesman problem may be of type P or it may not; at the moment it is not known for certain. What is known is that the problem is of type NP, which stands for 'non-deterministic polynomial time'. To understand this notion, imagine a Turing machine (or any other computation device) which is capable of making random guesses at various stages during its operation. (This has to be imagined, as there is no possibility of building such a machine.) By using such a hypothetical device (it is called a *non-deterministic Turing machine*), the travelling-salesman problem can be solved in polynomial time. The algorithm is simple. Guess the first location to be visited, then the second, then the third, and so on, until the entire tour has been guessed. Calculate the total distance involved in the tour and compare this with the given number B. Provided the machine 'guesses right' at each stage (in reality an unlikely event having probability $1/N!$, where N is the number of locations to be visited), the result obtained will be correct. This is what it means for a problem to be of type NP: by making one or more 'correct' (or 'optimal') guesses it is possible for a non-deterministic Turing machine to solve the problem in polynomial time.

Another problem that is of type NP is the testing of integers to see if they are composite (i.e. not prime). To test a given integer n, guess integers a and b less than n and check if $ab = n$. This will clearly give an answer in polynomial time, and an optimal guess will give the correct answer. But notice that the same approach will not be sufficient to establish that the 'complementary problem' of determining whether a given integer n is prime is of type NP. A single good guess is all that is required to show that n is composite, whereas *all* guesses have to fail in order to prove n is prime. In fact, primality testing is of type NP, but to demonstrate this you have to use a different method of primality testing.

The importance of this highly abstract concept of NP-type problems comes from a combination of two factors. Firstly, many of the problems for which no efficient algorithm has yet been discovered turn out to be of type NP. (Intuitively we can see that this is because the difficulty in such problems arises not from the necessary computational procedure, but solely by virtue of the vast number of possibilities. When all these different possibilities are sufficiently alike to be handled in the same manner, they are amenable to the guessing strategy that lies behind the NP concept.) Thus NP provides a theoretical framework for handling a great many real, practical problems.

The second factor stems from Stephen Cook's 1971 work on algorithm efficiency. By utilizing techniques going back to Turing and others, Cook was able to provide a means of demonstrating that certain NP-type problems are highly unlikely to be solvable by an efficient, polynomial-time algorithm. (That phrase 'highly unlikely' might in fact be replaceable by 'definitely unable', as we shall see.) Specifically, what Cook did was to prove that a particular problem of type NP was what he called *NP-complete*. What this means is that if this particular problem could be solved by a polynomial-time algorithm, then so too could every other problem of type NP. In other words, the problem considered by Cook is 'just as hard' as

Box C: Some NP-complete problems

The Travelling-Salesman Problem. (See the text for details.)

The Hamiltonian circuit problem. Given a network of cities and roads linking them, is there a route that starts and finishes at the same city and visits every other city exactly once?

The multiprocessor scheduling problem. Given a collection T of tasks to be performed, together with a list of the times taken to perform each task on a certain type of processor, and given also a specific number of processors of that type, is it possible to split up the tasks in T and assign each group to one processor so that the total time taken to perform all the tasks is less than some specified time? (Each processor works sequentially, though the entire collection of processors run concurrently.)

The map-colouring problem. (See chapter 7 for the background.) Given a map, is it possible to colour that map using just three colours in such a way that no two countries with a common border are coloured the same?

The quadratic residue problem. Given positive integers a, b, c, with a less than b, is there a positive integer x less than c such that

$$x^2 \bmod b = a?$$

Quadratic Diophantine equations. (See Chapter 6 for the background.) Given positive integers a, b, c, do there exist positive integers x and y such that

$$ax^2 + by = c?$$

every other problem of type NP. Capitalizing on Cook's result, a number of other mathematicians subsequently proved that a great many other NP problems are also NP-complete – including the travelling-salesman problem (see Box C).

So, as a result of the work by Cook and others, there is a way of demonstrating that a great many of the (type NP) problems that arise in the real world are as hard to solve as any other problem of type NP. Now, most mathematicians would conclude that it is a waste of time trying to find an efficient (i.e. polynomial-time) algorithm to solve a problem known to be as hard as every other NP problem. Hence a proof that a given problem is NP-complete is inevitably regarded as a proof that it cannot be solved by a polynomial-time algorithm.

But there is a difficulty here. Cook's result (and all the subsequent ones to date) does not preclude the possibility that the classes P and NP are in fact the same, i.e. that any problem of type NP is in fact solvable using a polynomial-time algorithm (though finding such an algorithm might be no easy matter in any specific instance). And if that were the case, knowing that a problem was 'as hard as any other in the class NP' would not really say very much. (All NP problems would be 'easy' in our present sense.) But few experts regard this as a likely possibility. The nature of the NP concept, involving as it does the highly non-algorithmic procedure of 'guessing' (in fact 'guessing right', usually against phenomenal odds), makes it improbable that it can be equivalent to type P. Consequently, the theoretical possibility that P and NP are the same is usually discounted at the outset, and an NP-completeness result *is* regarded as a proof that the problem is genuinely 'unsolvable'.

Of course, all that it would require to settle the issue conclusively would be to find just one problem which is of type NP but demonstrably not of type P. But despite the intuitive gap between the two notions P and NP, this has so far not been achieved, and indeed all the available evidence suggests that it is an extremely hard problem. It is known as the 'P = NP problem', and it is, as you might imagine, regarded as one of the most significant open problems of present-day computational mathematics. Part of its significance stems from its relevance to a great many practical problems, of course. But here one has to be cautious, for questions of relevance are never simple. So, it is time we returned:

Back to Real Life – Linear Programming

Though they can provide some valuable information, the theoretical techniques just described do not always give an altogether accurate picture of whatever it is to which they are applied. An algorithm may in theory run in exponential time (i.e. 'inefficiently') and yet, in practice – with everyday data, it may in fact perform very well. The exponential running may arise only for certain kinds of data not generally met with; to some extent the travelling-salesman problem falls into this category. When applied to 'real-life' configurations of cities and roads, the available methods – which are without doubt exponential in their running time – can perform quite well. An even more striking example of the potential gulf between theory and practice is provided by the so-called linear-programming problem, the problem which gave rise to (and continues to be one of the central themes of) the subject known as operations research. (This subject, whose origins lie in the Second World War, uses mathematical methods to assist in complex problems involving the direction and management of large systems of men, machines, materials, and money in industry, business, government, and defence.)

Linear programming is a technique used to provide a mathematical description (or *model*) of a real-life problem in which something needs to be *maximized* (e.g. profits or security) or *minimized* (e.g. costs or risks). The required *optimization* is achieved by a suitable choice of the values of a number of *parameters* (or *variables*). Both the factor to be optimized and some or all of the parameters will be subject to one or more *constraints*. The word 'linear' in linear programming indicates that all the mathematical expressions in the model are linear (i.e. do not involve the multiplication together of two or more variables or the raising of a variable to a power). In practice this is not a great restriction, since most of the problems encountered in real life either are intrinsically linear or can be assumed to be linear without giving rise to any great errors.

An elementary consideration of the problem will show that the linear constraints have a natural geometrical representation. Values of the variables which satisfy all the constraints correspond to points which lie inside a certain geometric figure. If there are two variables, that figure will be a polygon (whose number of sides corresponds to the number of con-

Colour of wool	Amount required per unit length		Amount available
	Cloth A	Cloth B	
Red	4 kg	4 kg	1400 kg
Green	6 kg	3 kg	1800 kg
Yellow	2 kg	6 kg	1800 kg

Table 5. Amounts of red, green, and yellow wool required to make unit lengths of cloths A and B, and the total amounts available.

straints), if there are three variables it will be a polyhedron, and if there are N variables it will be a polytope in N-dimensional space (see Chapter 10). Of course, there is no possibility of being able to draw a polytope in four or more dimensions, but the mathematics is straightforward whatever the dimension.

A simple example should serve to make this clear. Imagine a company which makes two kinds of cloth, A and B, using three different colours of wool. The amounts of wool required to make a unit length of each type of cloth, and the total amount of wool of each colour available, are given in Table 5. The manufacturer's profit is £12 per unit length on cloth A, and £8 per unit length on cloth B. The question is, how should the available wool be used to make the largest possible overall profit?

We begin by letting x and y denote the number of units of cloth A and cloth B, respectively, that are produced. This will yield a profit P (in £) of

$$P = 12x + 8y. \tag{10}$$

What are the constraints on the values of x and y? Well, since there is only 1400 kg of red wool available, and both types of cloth require 4 kg of red wool for each unit length, then

$$4x + 4y \leqslant 1400. \tag{11}$$

Similarly, by considering the available green and yellow wool,

$$6x + 3y \leqslant 1800, \qquad 2x + 6y \leqslant 1800. \tag{12}$$

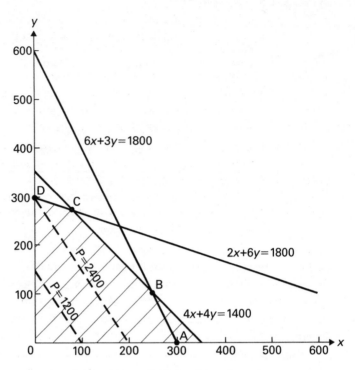

Figure 64. Linear programming. The solution to the cloth-manufacturing problem (see the text for details).

Finally, since neither x nor y should be negative (a constraint which is obvious when the actual problem is considered, but which has to be made explicit in the mathematical representation), there are the requirements that

$$x \geqslant 0, \qquad y \geqslant 0. \tag{13}$$

Figure 64 provides a graphical representation of the constraints imposed by the inequalities (11), (12), and (13). Any pair of values of x and y which satisfies all these constraints will be the coordinates of a point within the shaded region, and conversely any point in this region will have coordinates which satisfy the inequalities (11), (12), and (13). (Check this yourself by reading off the coordinates of various points from inside and outside the region.) So, what we have to do now is find a point within the shaded area which makes the quantity P in Equation (10) as large as possible.

Now, all lines whose equations have the same form as Equation (10) for a fixed value of P are parallel to each other. (Two such lines, for $P = 1200$ and $P = 2400$, are shown in Figure 64.) So it is fairly clear what we must do in order to maximize P: move the profit line (given by Equation (10)) as far away from the origin as possible without completely leaving the shaded region. This takes us to the point marked B. The coordinates of B are easily obtained by elementary algebra (as the solution of two simultaneous equations): they are $(250, 100)$. Thus the manufacturer must make 250 units of cloth A and 100 units of cloth B in order to obtain the maximum possible profit, which is £3800. (As it happens, this will use all of the available red and green wool, but leave 700 kg of yellow over, so our manufacturer should have done his homework before buying his wool.)

Having solved the problem, notice now what was involved. The constraints were represented in Figure 64 as the polygonal region ABCDO of the plane. The maximization point was one of the corners of the polygon – the remaining task was to find which corner. In this simple example, finding which corner presented no difficulty, but this is the part that makes more complicated (and hence more realistic) linear-programming problems hard. In a problem with three variables, the constraints will give rise to a three-dimensional polyhedron; with N variables you get an N-dimensional polytope, which cannot be drawn but which can still be handled algebraically. In every case what the problem boils down to is finding the vertex of the constraint figure (polygon, polyhedron, or polytope) where optimization occurs. But how do you do this? There could be millions (or more) corners, so an exhaustive search is usually out of the question, just as with the travelling-salesman problem. Some systematic way of flushing out the optimal corner is required.

In 1947, the American mathematician George Dantzig devised such a method: the *simplex algorithm*. In essence, what this method does is to start from one vertex (exactly how this initial vertex is found will not be described here) and then move around the surface of the polytope, following edges from vertex to vertex. Each time a new vertex is reached, there will be several (perhaps many) different ways of proceeding next, and there are various criteria for deciding which to follow (the most 'obvious' of which is to move only to a vertex which increases the quantity to be maximized – or decreases it if it is to be minimized).

Because of the enormous number of possible paths around the edges of a polytope, the simplex algorithm is known to be, in theory, an exponential-time algorithm, but when used in practice (on problems involving hundreds or even thousands of variables) it works extremely well, homing in on the optimal vertex in a relatively small number of steps. Indeed, the indications are that it tends to run in linear time (i.e. the number of steps is directly proportional to the number of variables involved). The types of constraints (and their associated polytopes) which cause the algorithm to operate inefficiently just do not arise very often in practice – they have to be specially 'cooked up' with the express purpose of defeating the simplex method. Indeed, so contrived are they that the existence of these 'artificial' problems does not for one moment prevent professionally implemented versions of the simplex algorithm being one of the first packages being made commercially available whenever a new computer system is put on the market for industrial and commercial use. Put simply, the method works.

But is there a faster method, one that is fast not just on most problems but in all cases – a polynomial-time algorithm? Intuitively, instead of finding the optimal polytope vertex by following edges around the surface of the polytope, it ought to be faster to 'take a short cut' across the interior. The difficulty with this approach is that, since you do not know in advance which vertex is optimal, how do you decide in which direction to proceed? By staying on the surface of the polytope at least you have a method for deciding which way to go at each stage. But is there any way of orienting yourself once you have 'cast loose' from the surface and headed into the interior?

In fact, there is. In 1970, the Soviet mathematician Shor realized that an old technique known as Newton's method could be applied to the linear-programming problem, and further modifications of this idea by Levin, Judin, and Nemirovski (also from the USSR) led to the formulation in 1976 of the so-called *ellipsoidal method*, in which the direction of the path to be followed across the interior of the polytope is determined with the aid of a sequence of ellipsoids drawn to 'approximate' the polytope. In 1979, Khachian (again from the USSR, which had a virtual monopoly of this technique) showed that the ellipsoidal method runs in polynomial time. Unfortunately, though this meant that the method was theoretically better than the simplex method, when applied to real-world problems it did not perform anything like as well as the simplex algorithm.

'So much for the theoreticians' concept of efficiency,' you might say, 'when the theoretically inefficient method easily outperforms the theoretically efficient one.' And many non-mathematicians did say just that. After all, the linear-programming problem was – and is – perhaps the single most important real-life problem. If there were to be any example to indicate a deficiency in the polynomial-time/exponential-time classification of efficiency, this was the worst one possible from the pure mathematicians' point of view.

But then, early in 1984, another theoretician came to the rescue. Narendra Karmarkar, a 28-year-old mathematician working for Bell Laboratories in the USA, discovered a polynomial-time linear-programming algorithm which really did work well in practice, even outperforming the simplex method to a remarkable degree on many occasions. (In one test with a 5000-variable problem, Karmarkar's algorithm was 50 times faster than the simplex algorithm.) It was a remarkable and quite unexpected advance, and to obtain his new algorithm Karmarkar had to make use of some highly sophisticated mathematics, involving a sequence of 'reshapings' of the polytope in order to obtain 'preferred directions' to follow once you are inside it. (Though, just as in computer implementations of the simplex algorithm, where the computer deals with arithmetic manipulations and not directly with the geometrical ideas behind them, so too when the Karmarkar algorithm is implemented the sophisticated geometrical ideas are suppressed in favour of a series of arithmetic operations on matrices.)

So the theoreticians' concept of efficiency was shown to work well after all. Moreover, the new algorithm provides a striking example of how some highly sophisticated abstract mathematics, involving multi-dimensional analogues of polyhedra and bizarre mathematical 'deformations', can lead to a concrete product of crucial importance in the 'real' world of business, commerce, and defence. Altogether an exemplary blend of the pure and abstract with the world we live in and an excellent place to bring to an end a survey of mathematics' New Golden Age.

Suggested Further Reading

A highly readable, brief account of algorithm efficiency can be found in the article 'The efficiency of algorithms', by Harry Lewis and Christos Papadimitriou, published in the magazine *Scientific American*, Volume 238 (January 1978), pp. 96–109. The standard introductory text for the study of algorithm efficiency is *Computers and Intractability*, by Michael Garey and David Johnson (W. H. Freeman, 1979).

For a low-level account of the linear-programming problem and the simplex algorithm, see the book *Newer Uses of Mathematics*, edited by James Lighthill (Penguin, 1978).

Karmarkar's algorithm is described in his paper 'A new polynomial-time algorithm for linear programming', published in the mathematical journal *Combinatorica*, Volume 4 (1984), Number 4, pp. 373–92.

Author Index

Subject Index

FOR THE BEST IN PAPERBACKS, LOOK FOR THE 🐧

In every corner of the world, on every subject under the sun, Penguin represents quality and variety – the very best in publishing today.

For complete information about books available from Penguin – including Puffins, Penguin Classics and Arkana – and how to order them, write to us at the appropriate address below. Please note that for copyright reasons the selection of books varies from country to country.

In the United Kingdom: Please write to *Dept E.P., Penguin Books Ltd, Harmondsworth, Middlesex, UB7 0DA.*

If you have any difficulty in obtaining a title, please send your order with the correct money, plus ten per cent for postage and packaging, to *PO Box No 11, West Drayton, Middlesex*

In the United States: Please write to *Dept BA, Penguin, 299 Murray Hill Parkway, East Rutherford, New Jersey 07073*

In Canada: Please write to *Penguin Books Canada Ltd, 2801 John Street, Markham, Ontario L3R 1B4*

In Australia: Please write to the *Marketing Department, Penguin Books Australia Ltd, P.O. Box 257, Ringwood, Victoria 3134*

In New Zealand: Please write to the *Marketing Department, Penguin Books (NZ) Ltd, Private Bag, Takapuna, Auckland 9*

In India: Please write to *Penguin Overseas Ltd, 706 Eros Apartments, 56 Nehru Place, New Delhi, 110019*

In the Netherlands: Please write to *Penguin Books Netherlands B.V., Postbus 195, NL–1380AD Weesp*

In West Germany: Please write to *Penguin Books Ltd, Friedrichstrasse 10–12, D–6000 Frankfurt/Main 1*

In Spain: Please write to *Longman Penguin España, Calle San Nicolas 15, E–28013 Madrid*

In Italy: Please write to *Penguin Italia s.r.l., Via Como 4, I-20096 Pioltello (Milano)*

In France: Please write to *Penguin Books Ltd, 39 Rue de Montmorency, F-75003 Paris*

In Japan: Please write to *Longman Penguin Japan Co Ltd, Yamaguchi Building, 2–12–9 Kanda Jimbocho, Chiyoda-Ku, Tokyo 101*

FOR THE BEST IN PAPERBACKS, LOOK FOR THE (🐧)

PENGUIN POLITICS AND SOCIAL SCIENCES

Political Ideas David Thomson (ed.)

From Machiavelli to Marx – a stimulating and informative introduction to the last 500 years of European political thinkers and political thought.

On Revolution Hannah Arendt

Arendt's classic analysis of a relatively recent political phenomenon examines the underlying principles common to all revolutions, and the evolution of revolutionary theory and practice. 'Never dull, enormously erudite, always imaginative' – *Sunday Times*

The Apartheid Handbook Roger Omond

The facts behind the headlines: the essential hard information about how apartheid actually works from day to day.

The Social Construction of Reality Peter Berger and Thomas Luckmann

Concerned with the sociology of 'everything that passes for knowledge in society' and particularly with that which passes for common sense, this is 'a serious, open-minded book, upon a serious subject' – *Listener*

The Care of the Self Michel Foucault
The History of Sexuality Vol 3

Foucault examines the transformation of sexual discourse from the Hellenistic to the Roman world in an inquiry which 'bristles with provocative insights into the tangled liaison of sex and self' – *The Times Higher Educational Supplement*

A Fate Worse than Debt Susan George

How did Third World countries accumulate a staggering trillion dollars' worth of debt? Who really shoulders the burden of reimbursement? How should we deal with the debt crisis? Susan George answers these questions with the solid evidence and verve familiar to readers of *How the Other Half Dies*.

FOR THE BEST IN PAPERBACKS, LOOK FOR THE 🐧

PENGUIN BUSINESS AND ECONOMICS

Almost Everyone's Guide to Economics
J. K. Galbraith and Nicole Salinger

This instructive and entertaining dialogue provides a step-by-step explanation of 'the state of economics in general and the reasons for its present failure in particular in simple, accurate language that everyone could understand and that a perverse few might conceivably enjoy'.

The Rise and Fall of Monetarism   David Smith

Now that even Conservatives have consigned monetarism to the scrapheap of history, David Smith draws out the unhappy lessons of a fundamentally flawed economic experiment, driven by a doctrine that for years had been regarded as outmoded and irrelevant.

Atlas of Management Thinking   Edward de Bono

This fascinating book provides a vital repertoire of non-verbal images that will help activate the right side of any manager's brain.

The Economist Economics   Rupert Pennant-Rea and Clive Crook

Based on a series of 'briefs' published in *The Economist*, this is a clear and accessible guide to the key issues of today's economics for the general reader.

Understanding Organizations   Charles B. Handy

Of practical as well as theoretical interest, this book shows how general concepts can help solve specific organizational problems.

The Winning Streak   Walter Goldsmith and David Clutterbuck

A brilliant analysis of what Britain's best-run and most successful companies have in common – a must for all managers.

Lateral Thinking for Management Edward de Bono

Creativity and lateral thinking can work together for managers in developing new products or ideas; Edward de Bono shows how.

Understanding the British Economy Peter Donaldson and John Farquhar

A comprehensive and well signposted tour of the British economy today; a sound introduction to elements of economic theory; and a balanced account of recent policies are provided by this bestselling text.

A Question of Economics Peter Donaldson

Twenty key issues – the City, trade unions, 'free market forces' and many others – are presented clearly and fully in this major book based on a television series.

The Economics of the Common Market Dennis Swann

From the CAP to the EMS, this internationally recognized book on the Common Market – now substantially revised – is essential reading in the run-up to 1992.

The Money Machine How the City Works Philip Coggan

How are the big deals made? Which are the institutions that really matter? What causes the pound to rise or interest rates to fall? This book provides clear and concise answers to these and many other money-related questions.

Parkinson's Law C. Northcote Parkinson

'Work expands so as to fill the time available for its completion': that law underlies this 'extraordinarily funny and witty book' (Stephen Potter in the *Sunday Times*) which also makes some painfully serious points about those in business or the Civil Service.